DEAD END PATH

DEAD END PATH:

How Industrial Agriculture Has Stolen Our Future

By David L. Brown

Moab BookWorks

A Small Press for a Small World

Introduction

This is not just a book about farming. It's an extended essay on how industrial agriculture has led us down an unsustainable path that threatens our very civilization and the future of the human race. The danger is real and looming before us in the here-and-now. Our petroleum-based machine technology is reaching its limits, which is likely to trigger a domino-like food security crisis that would destroy our way of life and change the world forever.

The book is the product of a lifetime of research, experience, study, conversation and contemplation. It carries an important message for our troubled Earth, one that I've probed and assessed for decades and to which I can bring insight from a range of viewpoints. The book is fact-filled and written to interest non-technical readers.

Let me introduce myself. I'm not an ivory tower academic. although I am familiar with those places. Nor am I a scientist, although I've spent my life as a lay student of science. I'm not a farmer, but I've met and written about thousands of them all across North America, Europe and the Pacific Rim.

I'm a journalist with roots in rural America, and I've had a unique opportunity to observe and report on a world that's reeling toward disaster. I've been part of the problem too, working for agri-business companies engaged in the spread of industrial agriculture, the villain of my story. This book emerged from my journey of discovery and study. In a way, it seems my entire life has been in preparation to create this essay on the state of civilization.

My story, like any good tale, begins at the beginning.

As a small child I spent the last years of World War Two on my grandfather's 40-acre farm in northern Indiana. Born in 1875, he was of Scots ancestry. Successful in business, in his early years he was active in politics, twice running as a candidate for Indiana Secretary of State. He made a fortune as a dealer in exotic furniture grade lumber, lost it all to the Great Depression, and returned to his farming roots during the 1930s. Much later in his late 70's he became the oldest American to that time to qualify for social security by paying into the system from earnings of his roadside vegetable stand.

Grandpa never owned a car or tractor and even into his 70s he worked his land with a team of draft horses. One of my earliest memories is of him with those great, broad-rumped beasts. He was a horseman to the end for I was told that as he died in his late 80's, in the fog of a terminal coma he shouted "Gee" and "Haw," revisiting in his imagination all the teams of his long life.

After the world war I grew up in central Missouri where my father was a professor of entomology at the University of Missouri. Besides teaching he worked on grants from the U.S. Department of Agriculture and corporations such as Monsanto to help develop and field test insecticides to kill corn earworms. The poisons, known as organic phosphates, were derived from nerve gas chemistry. Applying them to pest control was one facet of the great burst of scientific farming that was called the Green Revolution. Visiting test sites with my father, I was a witness to that.

We lived in the country on what today would be called a

"hobby farm," about eight acres where we maintained a milk cow and at times pigs and even goats plus chickens, ducks, geese, rabbits, and pigeons. Our pond was stocked with largemouth bass and bluegills.

While others played or traded baseball cards, my childhood was largely devoted to helping maintain a highly-diversified farm writ small. It was my job to bring in the cows from the pasture, tote bales of hay to feed them, muck out the barn, gather eggs and perform many other chores.

We maintained extensive gardens where I became adept at weeding, hoeing and spreading compost. I also had the tasks of pruning and spraying a small orchard of perhaps 50 trees yielding peaches, apples, pears and cherries. We grew strawberries, raspberries, grapes and melons. Our produce was stored for the winter in a walk-in cooler and large freezer. Already handy with saws and hammers, at age 13 I did much of the work in building a new pole barn with milk stalls for two cows and a large haymow.

For grades two through six I attended a one-room country school where nearly all my fellow pupils were farm kids. I belonged to 4-H and won blue ribbons at the county fair for my Jersey heifer and other projects.

Those years in the early 1950s were simpler times, and yet the dangers that faced the planet were already being discussed at our dinner table. Despite the promise of science to increase food production, my father was convinced that over-population would lead to disaster. His thinking was perhaps ahead of its time, for it would be many years before Paul Ehrlich, a fellow entomologist, would write *The Population Bomb*.

My journey of discovery continued. I earned a degree in journalism from the University of Missouri, worked as a newspaper reporter then joined the staff of a farm magazine as a reporter-photographer. From there I moved to Chicago and a job as a writer at International Harvester Co., a major agri-business corporation which then ranked No. 19 on the Fortune 500 list. I traveled extensively across America, observing how farm equipment was marketed and used.

After leaving that job I took a two-year hiatus from agri-culture, working as publications director for a major health charity, the Easter Seal Society. Then, perhaps reflecting my grandfather's entrepreneurial instincts, I launched on my own as a freelance writer and photographer. For a while I did general coverage in Chicago, stringing for medical and industrial publications and covering such events as the 1968 Democratic National Con-vention. But my roots in the field of agriculture drew me back and soon I was specializing in work for farm publications and agri-business clients.

For two years I traveled the U.S. and Canada on freelance assignments, interviewing farmers, making videotapes and stills, writing articles for farm magazines. In 1971 I organized, promoted and led a series of farm tours of Europe, drawing several hundred North American farmers and spouses to visit farms in eight nations.

After that, with a few dollars in my pocket, I formed an editorial services agency, producing a wide range of magazines, videos, newsletters, speeches and other marketing tools for agri-business clients. Over the years I hired, trained and managed more than 20 writer-photographers, nearly all of them with farm backgrounds and degrees in agricultural journalism.

One of the largest of my clients was A. O. Smith Harvestore Products, Inc., a maker of glass-lined silos, automated feeding equipment and manure handling systems for livestock and dairy farmers. For two decades I edited and produced *Harvestore Farming*, a magazine featuring users of the company's products. During that

time I continued to visit and write about cutting-edge farm oper–
ations and interview agricultural scientists at leading universities.
At one point I wrote a farm management handbook, a complete
guide to livestock and dairy farming.

During those years and ever since I maintained an interest in a
variety of subjects related to the thread of this book, including
economics, science and technology, and especially everything to do
with the environment. In 2006 I began a weblog, Star Phoenix
Base (www.starphoenixbase.com), in which I posted essays on a
number of the subjects that interest me. The title came from a
novel I wrote during the 1990s about a future when over-
population has brought down civilization. The blog is identified as:
"Your portal to the world of the near future and the dangers that
lie ahead." Many of the ideas expressed in this volume were first
explored and developed in the several hundred essays I wrote over
this four-year extended phase of my journey of enlightenment.

For many years I've been a member of the American
Association for the Advancement of Science and the American
Agricultural Editors Association, and am 50+ year member of the
Society of Professional Journalists, which I recently served as
president of the New Mexico chapter.

And, always since I was five years old, I read. My house is
filled with thousands of volumes. When the idea for the present
work began to stir in the back of my mind I revisited some of those
books, re-reading and making margin notes. I began to obtain
other volumes, some current and others out-of-print but available
from used book dealers through Amazon. I searched the Internet
and printed out reams of articles, research reports and news stories.
I tore out and filed hundreds of pages from *Science*, *The Economist*
and others. In the end I had gathered nearly 150 volumes and
several file drawers filled with reference material.

The concepts expressed in the following pages emerged naturally out of all those years of real-world experience, study, research and thought. The book is in essay style. I've combined the skills of the journalist and the power of my research, seasoning the story with personal observations, experiences and opinions, both my own and those of others.

I've tried to write in a conversational way, as if I were sitting across from my reader, whom I assume to be intelligent, curious and appreciative of wit. I sometimes use analogy and metaphor to make important points, my aim being to make the read interesting and entertaining as well as informative. And finally, although my subject is as serious as a heart attack, I've tried to avoid projecting a sense of unremitting gloom. The facts are ominous enough.

I've added extensive footnotes, something I've always favored not only as a way for writers to list references, but also to make supplemental comments. I've freely exercised the latter option, sometimes in a whimsical way intended to bring a smile to the reader's lips. And, oh yes, this book has an extensive bibliography to serve as a useful reference and guide to further study for the reader.

I've led an interesting life, seeing and learning many things from different vantage points. My observations and conclusions about the ominous course upon which the world is set are solidly based on facts, science, and the opinions of others far more learned than I.

My goal in writing Dead End Path is to make a new cadre of readers aware of the dangers and challenges that lie ahead. You may not agree with all I have to say, but I hope my words will stimulate your curiosity and lead you to think deep and long about what is undoubtedly the most important, yet often overlooked subject of our time. — *David L. Brown*

Chapter One

The Unusual Suspect

Open your mind and look around and you will see a world in distress. Each day seems to bring news of fresh disasters. The media spill out an endless stream of stories about growing threats to the environment, troubled economies, and dangers to the climate. We are overwhelmed by reports of failing nations, crime, intolerance, disease, terrorism, piracy, war, natural disasters, drought and spreading famine.

What's happening to bring so much misery into the world? The subject may seem complex and multi-faceted, but there's a single culprit responsible for most of the troubles we face — an "unusual suspect" that is the root cause of nearly everything that's wrong on Planet Earth.

Aha, you may be thinking: If there's a single cause for most of our troubles it should be easy to solve them, simply by eliminating that cause. But you would have to think again when I reveal that the culprit is agriculture, and in particular the aggressive style of industrial farming that has replaced traditional methods in developed countries and is being spread across the world by multinational corporations through the process of globalization.

But … isn't agriculture a force for good? It puts food on all our tables. Its bounty makes possible the comfortable lifestyles most Westerners enjoy. It feeds the less fortunate. It allowed our species to thrive and build great civilizations. Agriculture lies at the heart of humanity's finest achievements.

Doesn't it?

In the short-term that may seem true—but just as modern agriculture has allowed us to create the world as it is today, it has also led us down a dangerous dead end path by creating conditions that cannot be sustained. As a result, our civilization is teetering on the brink of disaster.

As the late economist Herbert Stein once wrote,[1] "If something cannot go on forever, it will stop." To put it bluntly, industrial food production cannot go on forever. At some point it must begin to stop. That time may already be upon us.

The Case Against Agriculture

Let us visit an imaginary courtroom with personifications of Nature[2] as judge and jury. Before the bench stands industrial agriculture, charged with high crimes against humanity. What evidence could the prosecution bring? As it happens, the facts are many and conclusive. This book will examine that body of evidence, laying out a case on which no informed judge or jury, much less Nature herself, could fail to convict.

---

[1] Herbert Stein, "Stein's Law," quoted in *Slate* magazine, May 16, 1997.

[2] Throughout this book I use the capitalized form to indicate the essential importance that Nature has on the fate of humanity. I do not mean to imply that she is a living, thinking being; this is merely metaphor. However, it would do humankind much good to embrace a philosophy that holds her in deepest respect. I refer to her in the feminine form because in truth she is the mother of us all.

I'm pointing out no individuals for blame, especially not farmers and ranchers. No less than the rest of us, and more than most, those who have remained closest to the land are victims of the juggernaut of technology and economic exploitation that has swept away traditional farming. It's the very process of industrialized agriculture that is the villain in this story. I suspect many agriculturists will agree with the ideas expressed here.

But how could agriculture, the critical process of providing food for our tables, have stolen the future of the human race? That's a pretty harsh claim, yet the evidence is clear. In essence, technological farming has broken the natural cycles of the environment through the conjunction of machinery and temporarily cheap and abundant resources. It has caused our species to unwisely set ourselves against the very environment of which we are an integral part.

Sometimes it seems as if we've literally declared war on our own planet. That idea is explicit in oft-heard phrases about "conquest," "triumph," or "victory" over Nature, as if our very environment is an enemy to be attacked and defeated.

Unfortunately for us, history reveals that in the end Nature always wins. If you doubt it, ponder the fates of the Akkadians, the Sumerians, the Hittites and the Babylonians whose early agricultural civilizations lie forgotten beneath the dusts of time.

It's impossible for us to "win" against our own planet, because the Earth can perfectly well continue without the burden of the human race. It needs us no more than a dog needs fleas. Nature is patient and her wheel turns slowly, but inexorably.

We've gone down a dead end path by presuming to step outside of Nature, to assume mastery over the very environment.

Like Goethe's sorcerer's apprentice[3] we have meddled with powers far beyond our control. Now we're beginning to reap ecological, social and cultural disaster as the price of that hubris.

In a classic example of the law of unintended consequences, industrialized farming has allowed human population to expand beyond natural limits. There are now well over a thousand times more people alive on Earth than at the dawn of agriculture when our ancestors lived close to Nature. According to many estimates we've already exceeded the carrying capacity of the Earth, and as resources dwindle there will be an enormous price to pay. There are too many of us "at the table," and more guests arrive each second of every day.

As industrialization takes us down the dead end path we're eating up the Earth's supply of irreplaceable natural resources. Many of these, particularly petroleum, are on the verge of serious decline. Without these essential inputs our present food production systems cannot be sustained.

It all started innocently enough about 10,000 years ago when humans set aside the tools of hunting and gathering to take up the sickle and the digging stick. Thus our ancestors became farmers and began to break the connection between humans and Nature.

For thousands of years men and women have worked the land. Although they didn't realize it, even from the start they were beginning a process of destruction.

During the last few generations, and particularly since the Second World War, the pace of that destruction has multiplied many times over, sending humanity pell-mell down the dead end path to an uncertain future.

---

[3] Wolfgang Goethe, "Der Zauberlehrling," ("The Sorcerer's Apprentice"), a poem, 1797.

The Green Revolution

It may surprise you to hear farming described as a threat to humanity. It's been almost universally presented as a force for good. Consider the so-called "Green Revolution,"[4] that heroic label for the rapid industrialization of agriculture that took place during the 20th Century, driven by profit-seeking corporations and aided by government subsidies, loans and food aid.

The Green Revolution is claimed to have "saved" hundreds of millions of lives in the Third World. That's a good thing — isn't it? Yet we are seldom reminded that many of the supposed "ben–eficiaries" of that technological "miracle" are today living hopeless lives of poverty on the razor's edge of famine.

Consider that without Western intervention — through the Green Revolution, medical vaccines, antibiotics and food aid — the forces of Nature would have prevented most of those indiv–iduals from being born. The fortunate members of a smaller world population would be far better off. Hundreds of millions have been "saved" only to live in hunger and without hope.

We must ask ourselves: What kind of "good" is it that can create so much evil?

Starting in the 1940s, applied science and technology allowed food production to drive unprecedented population growth for

---

[4] The "Green Revolution" is a term applied to a worldwide burst of food production after the end of World War II. During the 1970s food surpluses actually became a problem, particularly in the U.S., which responded by exporting huge quantities of grain and other foodstuffs to poor countries as foreign aid (and not coincidentally, to help improve the lot of American farmers by subsidizing continued production). It should be realized that the Green Revolution was merely a new, more virulent stage in the process of farm industrialization. It did not provide a permanent solution, but only a temporary "fix" that put off the day of reckoning while allowing another jump in world population. We will examine this subject in Chapter 4.

over half a century, more than tripling human numbers. Now food supplies are beginning to falter even as population continues to rise. By 2009 there was evidence that food production was peaking or actually in decline. The UN reported that more than a billion people were malnourished, and that the number was rising.[5] These are the "fortunate" clients of the Green Revolution.

The Rush Toward ... What?

It is remarkable that during a brief period of history, less than a single lifetime, aggressive forms of technological agriculture have sent us hurtling down the dead end path toward that point when ... it soon must stop. Unless something is done, civilization may be in danger of falling into an abyss that would make the Dark Ages look like a sunny afternoon in June.

In its purest forms, industrial agriculture uses the face of our planet as a sort of Petri dish, a mere substrate on which genetically modified seeds, quantities of artificial fertilizer, and chemicals are mechanically applied to soil that has been robbed of its natural richness. Large yields of crops are produced, but with little regard to the health and future survival of the planet itself, much less to the well being of future generations of humans.

When not being ruined through aggressive mechanical and chemical farming, formerly rich lands are under continued attack by the construction of houses, office buildings, shopping centers, roads, airports and parking lots.

Extensive irrigation is steadily poisoning millions of acres with mineral salts. Finally, development and mismanagement causes the loss of millions of tons of precious topsoil each year to wind and water erosion.

---

[5] "More people than ever are victims of hunger," press release, Food and Agriculture Organization (FAO), United Nations, Rome, 2008.

The problem is that the resources used to achieve this are in limited supply and are being used up. When those resources are no longer readily available and affordable, how will we continue to feed a growing population of hungry human beings?

Most importantly, this unnatural form of agriculture has been made possible by cheap and plentiful energy, mostly from oil and natural gas used for gasoline and diesel fuel and as a feedstock for ag chemicals and fertilizer, not to mention powering the entire web of mining, manufacturing, processing, and transportation that is the underpinning of the world's food economy.

Without cheap and plentiful oil, there is no cheap and plentiful food. Ominously, the Earth's supply of oil is falling behind demand and there is no prospect for a practical solution to the looming energy crisis. From this we can conclude that a growing "food gap" lies before us. That poses ominous consequences, especially for those the Green Revolution "saved" and who are already living in the looming shadow of outright famine.

It gets worse, for the rampant burning of fossil fuels is warming our planet, causing climate change that is certain to affect our food security through drought, floods, heat and storms. There's ample evidence that such changes are already taking place. The industrialization of China and India through increased use of coal is driving this to new heights.

Even water, another precious natural resource essential to the continuation of industrial farming, is reaching its limits. Fresh water is a direct analog of farm-produced food, for without it there would be no crops to harvest. From the river basins of Asia, to Australia, Europe, and the American West, water resources are dwindling. Glaciers are melting and many of the world's major rivers sometimes fail to reach the ocean because all their waters are used up for irrigation and other industrial purposes.

It's not only the land that is being despoiled. Only a few decades ago we were reassured by predictions that the "bounty of the sea" would soon feed the world. That dream has turned to bust. Fisheries everywhere are threatened or collapsing as a result of aggressive over-harvesting, pollution, the creation of "dead zones" by fertilizer runoff and acidification by CO_2 absorption.

Myopic and sometimes bizarre economic theories also play a part in the case against technological agriculture. Inexplicably, most economists fail to take into account the long-term effects of resource depletion and all-out industrial food production. Most models of what is called neoclassical economics measure economic success and "progress" in terms of GDP growth, which is eventually unsustainable since it is based on the "development" (read: "using up" or "destruction") of irreplaceable resources.

Can it be true that economies can "succeed" only by traveling down a dead end path toward an unsustainable future? That seems extremely unlikely, but it will take more than just a little attitude adjustment for humanity to start thinking in a different way. The problem is that there is too much focus on short-term profits, too little on moral obligations to future generations.

Some argue that it doesn't matter whether we use up the resources of the Earth because our problems can be solved through human expansion into space. This is a false argument. There are many good reasons why future populations cannot take up house-keeping on transformed Mars or Venus. Mars is sterile and nearly airless, while Venus is smothered in a blanket of CO_2 that sustains an ambient temperature of around 460° C (860° F). The resources required to transform those alien worlds into second Earths is beyond imagination.

Nor does it seem likely that humans can live permanently in floating habitats in space or colonize undiscovered and far distant planets circling other stars. It's frankly impossible for the human race to relieve population pressure through off-world immigration because the cost would be literally astronomical. Interplanetary or interstellar solutions are merely pipe dreams of the self-deluded.

Because human beings are part of an almost infinitely complex web of life evolved over several billion years, the Earth is the only home we humans will ever have.

No matter how much we would wish it to be otherwise, the dead end path leads to a not-far distant time when unsustainable industrial agriculture will stop, and with it our present civilization.

That's a daunting prospect for our future. What options do we have? What steps could we take now? What unseen dangers lie ahead? Clearly, if our species is to have a future at all we must create new and much different forms of civilization, ones that place a far lighter "footprint" on the Earth.

The Future Must Be Sustainable

In the long term, societies must become completely sustainable, relying entirely on resources that are renewable or those that are in essentially infinite supply such as stone, wood, plant fibers, sand, clay, animal products, sunlight, water and wind. We must eventually learn to do without all non-renewable resources, including oil, gas, coal, iron, copper, lead, tin, uranium—in fact, virtually every mineral substance on which we presently rely.

Unfortunately, no longer-term solutions can be imagined that do not require an eventual reduction of human population. At some point, we must not only slow down but dramatically reverse the growth in human numbers. No eventual future can be imagined without true sustainability, and the Earth can continue to support only far smaller numbers than are present today, and certainly not in a time after nonrenewable resources have been depleted. Only when our descendants are in balance with renewable natural resources and populations remain stable or shrinking can our species continue to exist.

We should take a moment to ponder the meaning of the word "sustainability." The concept lies at the heart of the problems that face humanity, yet the term is commonly misused. Too often it's applied only to indicate that a certain practice is less damaging to the environment than others. For example, we may be told that no-till farming using chemicals is sustainable because it reduces soil erosion compared with regular plowing and disking. That misses the point because the soil is still being exploited and drained of its nutrients. The practice only pushes back the day when the piper must be paid.

Of course, nothing lasts forever. In a few billion years the Sun will expand as a red giant and engulf the Earth. No species can be expected to endure for anything like that span of time. The question we should consider is: How long does the human race wish to continue to exist? To define "sustainability" we must describe the actions necessary to meet that target. At our present rates of population growth and resource depletion, the time remaining is dangerously short.

Any practice that uses up irreplaceable resources—no matter how slowly—is unsustainable. By that rule our entire industrial civilization is by definition traveling down that metaphorical dead end path to an uncertain future.

Modern humans are Johnnies-come-lately on the scale of geological time, less than a few tens of thousands of years. By comparison the dinosaurs ruled for a hundred million years or more—several thousand times longer than our species has existed. They were able to do so because they had no ability to control Nature, and thus were forced to live within her boundaries. That's the key to sustainability in its true meaning. If human beings are to continue to live on Earth our descendants must return to a state of balance with Nature. There can be no more cheating, such as robbing the Earth of its oil and minerals. Fortunately, for Nature, that question is moot because in the near future there will be nothing left to steal.

Giving up our unsustainable civilization poses a significant challenge because nearly all of our present ways are diametrically opposed to the principles of sustainability. We are to be left in the position of the iconic man crawling across a sandy dune with a trackless desert all around, vultures circling above. As the Sun beats relentlessly down he takes the last sips of water from his canteen. For him, "sustainability" is possible only for as long as the water holds out, and the canteen is nearly empty.

It's not impossible to begin to move away from the unsustainable world we have created for ourselves, but it is surely the most immense challenge ever faced. We need to act soon or suffer the consequences. It will not be an option because our industrial food production is reaching its limits and is already straining to support present numbers. As resource peaks are passed and supplies begin to fail, the dangers will multiply exponentially.

How will the human race make this difficult but necessary transition? Will our species even survive?[6] These are serious questions. In the distant past our ancestors earned their livings from a close relationship with Nature. In the distant future there is no doubt that our descendants, if any, will by necessity have returned to some forms of natural existence.

---

[6] Some scientists believe the reason for the failure thus far of SETI, the Search for Extraterrestrial Intelligence, may be because interstellar civilizations soon collapse due to resource depletion, leading either to extinction or a non-technological lifestyle. Indeed, in a de-industrialized future it's hard to imagine that our descendants will be signaling to the stars.

In between, to quote T.S. Eliot, "lies the shadow,"[7] preventing us from seeing a clear picture of how our descendants may find their way to sustainability. Total disaster for our species may not be inevitable. There are many possibilities and although we cannot know how things will play out, it's interesting and perhaps essential for us to speculate.

According to the creation myths of many early cultures, the first humans were given stewardship over the planet and all it contains. Too often that was taken as a license to plunder the resources of the Earth. In light of present knowledge, no reasonable philosophy or religious teaching could approve of that misguided interpretation. The concept of stewardship must be turned on its head and viewed as a responsibility and a duty to our planet.

As a species we have acted with far more cleverness than wisdom. The challenge for our descendants will be to cultivate common sense solutions while making peace with Nature. It's imperative that our descendants become attuned with the environment, learning to live and work within the ecology of the Earth, husbanding its vast potential rather than destroying it for immediate gain.

In short, our challenge is to invent new and better ways of "being human." In the long stretch of future possibilities, true and beneficial stewardship of the Earth must become the primary essence of any successful human society. If we fail to achieve that, like the dodo or the passenger pigeon our species will simply cease to exist.

---

[7] T.S. Eliot, "The Hollow Men," a poem, 1925.

Chapter Two

Places at the Table

Probe deeply enough into any given problem facing human civilization and you'll almost invariably be able to trace the cause to overpopulation. The excess of human numbers is the uncomfortable force that is shaping the direction of our civilization, the looming problem that too few choose to recognize.

Follow the trail of evidence one step further back and it leads to industrial agriculture, the root cause of overpopulation.

According to an estimate by the late Paul MacCready, human beings and our domesticated livestock and pets may account for 98 percent of the total mass of vertebrate life on land.[8] Think of it! If he was right, only two percent is accounted for by all the wild animals and birds, including everything from armadillos to zebras, from eagles to hummingbirds, from wee colonies of mice to the elephants of the African veldt, and from endangered tigers prow-

---

[8] *Interview with Paul B. MacCready*, California Institute of Technology Oral History Project. Dr. MacCready (1925-2007) was an environmentally conscious aeronautical engineer who designed and built the world's first solar-powered airplane, the Gossamer Condor.

ling equally endangered jungles to the doomed polar bears of the rapidly thawing North.

Besides a human population of about 7.3 billion as of mid-2015, there are about 1.3 billion cattle, 2 billion hogs, 1.2 billion sheep, and 16 billion chickens alive on Earth at any one time. By comparison, African elephants number about 60,000. There are about 20,000 polar bears and their numbers may be declining. There are about 5000 tigers left in the world and three out of eight sub-species have become extinct since the 1940s.

Put this in sharper focus with the following factoids: There are about 33,000 pigs for every African elephant. There are about 65,000 cows for each polar bear. And the sheep-to-tiger ratio is about 240,000 to 1. With figures like these it's easy to see how Dr. MacCready came up with his statement.

The imperative to reproduce is in our genes. But like all other species, our ancestors evolved in such a way that through most of prehistory human numbers generally remained at peace with Nature. Early humans were part of an evolving ecology that had existed for hundreds of millions of years, and as long as they lived within Nature, not as its would-be masters, the balance took care of itself.

A few hundred generations ago, a mere tick in geological time, something changed. Our numbers began to grow at an unpre–cedented speed. Evidence is mounting that human beings are literally outgrowing the Earth.

What triggered this change? The obvious suspect is agricul-ture. It was when humans switched from a lifestyle as hunter-gatherers to farming and herding that population began to climb. And since the advent of industrial-scale agriculture in the last few decades, that rate of increase has shifted into overdrive.

Let's put this into perspective. Recent findings point to the earliest modern humans in what is now Ethiopia about 160,000 years ago.[9] For the sake of comparison, let's set that time 1600 centuries ago as our starting point for a historical review of population.

For 150,000 years before the appearance of primitive agriculture our ancestors were hunter-gatherers, living off the land just as other animals had done for millions and even a billion or more years. Because population numbers remained virtually static for most of that time, we can surmise that the number of people who existed on Earth was in some way related to the balance of Nature. After thousands of generations, by the dawn of agriculture the planet's population may have stood at around five million— about the number living in the present-day nation of Singapore.[10]

Over the vast span of 150,000 years, human numbers increased very slowly. The rate of increase at a given time, if any, was glacial. Here's a simple thought experiment to put the speed of growth into perspective. Let's assume that the population started at zero just before the first genetically modern human appeared in Africa, and grew steadily over a span of 150,000 years to reach five million by the dawn of agriculture 10,000 years ago.

We use our handy electronic abacus to determine that a population of five million divided by 150,000 years yields an average net rate of increase of just 33.3 persons per year. The population of the entire world increased at an annual rate so close to zero that there is hardly a difference.

---

[9] *"160,000-year-old fossilized skulls uncovered in Ethiopia are oldest anatomically modern humans,"* news release, U. of California Berkeley, 11 June, 2003.

[10] Paul R. Ehrlich and Anne H. Ehrlich, *"Population, Resources, Environment,"* Freeman Publishing Co., 1970. Other estimates of prehistoric population before the dawn of agriculture vary, ranging from as few as 1 million to as many as 10 million.

In fact, setting aside the unlikely assumption of a steady, straight-line increase, human numbers could have risen relatively quickly (in terms of geological time), then remained near to a balance point, which may have been around the number present at the threshold of the agricultural age. Conversely, they may have remained at a lower level for most of that time, maintaining a status quo. We just don't know the details, but we can conclude that when living in sync with Nature, as was the case prior to the invention of agriculture, human numbers were constrained.

But then something happened. If we look at the period starting 10,000 years ago, the picture changes dramatically. By 2000 B.C. there were about 27 million people on the planet. A thousand years later, around the time of the Trojan War, the figure had reached 50 million.[11] At the beginning of the Christian Era there were an estimated 200 million humans living on Earth—40 times more than at the end of the hunter-gatherer period.

It took 150,000 years for human numbers to reach five million, and in just a fraction of that time the human population had taken a huge jump. What was going on?

It appears there is some significance to that time 10,000 years ago when human numbers suddenly began to soar—and indeed there is. As we have seen, it was about then that our ancestors began to give up the life of hunting and gathering, of living close to the natural environment of the Earth. Some of our hunter ancestors transformed themselves into herders of sheep, cattle and goats. Former gatherers began to domesticate and cultivate wheat, rice, corn and many other plants.

---

[11] Due to uncertainties there are various estimates of early populations, particularly in prehistoric times. The figures used in these comparisons from 2000 B.C. through 1950 are from "*Atlas of World Population History*," Colin McEvedy and Richard Jones, 1978, Facts on File.

Dawn of the Age of Agriculture

The human species, always too clever by far, had invented something new. The first stirrings of what we might call the Age of Agriculture had arrived, and with that change the link with Nature was broken. Humans no longer lived within Nature's constraints, surviving on what she provided.[12] Rather than existing in tune with the ecology of the planet—an almost infinitely complex web of life that was the product of more than three billion years of evolution—the new herders and farmers presumed to step outside of Nature and become her masters.

Farming appears to have sprung up at almost the same time in many places, most famously in the Fertile Crescent of present day Iraq but also in Southeast Asia, Oceania, South and Central America, and elsewhere. Human beings no longer existed at the whim of their environments, for they were learning to exploit the resources of the Earth. And, pertinent to our discussion, through agriculture they obtained the means to feed more people.

Let's take a closer look at the population increase from 5 to 200 million that took place in the 8000 years prior to the beginning of the Christian Era. Using our trusty electronic abacus we divide 195 million additional people by 8000 years to learn that during that period the world population grew at a median rate of 24,375 people per year. Compare that with the scant increase of merely 33.3 per year in the pre-agricultural era. In terms of absolute numbers, population grew more than 700 times faster.

---

[12] In fairness it must be noted that humans had likely begun to have an impact on the environment prior to agriculture. Archaeologists have found that even proto-human *homo erectus* had mastered fire more than three quarters of a million years ago. Early hunters are known to have used fire to drive game and create open pastures. That may have had some impact, but fire is part of the natural cycle of nature so human activities were far less damaging to the environment than what is happening in our present time.

But that's nothing compared with what happened since the emergence of industrial agriculture. Fast forward again to recent times and we see that an incredible population surge has taken place. By 1750, just prior to the American Revolution, the world's population stood at about 750 million. Since the time of Christ the population had grown by about 550 million, a median annual rate of increase of more than 300,000.

Jump another century to 1850. At that time there were 1.2 billion people living on Earth—and the first steps were being taken toward what grew to become industrial agriculture. Eli Whitney introduced the cotton gin in 1793, leading to a huge spurt in cotton production across the South. Inventor Cyrus McCormick demonstrated his mechanical reaper in 1831. Six years later a blacksmith named John Deere invented the self-cleaning plow, making it possible for pioneers to break the heavy sod of the Midwest.

The back-breaking labor always associated with traditional farming was beginning to be replaced by the methods of the new Industrial Revolution. Soon to follow were the first steam tractors, gang plows, spring-tooth harrows, automated twine binders, horse drawn combines and a host of other mechanical devices. By the beginning of the 20th Century, industrial methods were well on the way to replacing traditional agriculture in the United States and other developed nations.

And with the advent of industrial agriculture, population growth leaped onto an even faster track. In the 100-year span from 1850 to 1950, human numbers doubled once again, reaching 2.4 billion. The median rate of increase had climbed to more than 12 million per year, 360,000 times higher than in pre-agricultural times.

But that crashing tidal wave of growing human numbers did

not stop there. Incredibly, in just 50 years by mid-2000 the world population increased by another 3.7 billion to 6.1 billion.[13]. In mid-2015 it stood at about 7.3 billion. Despite falling birthrates in some areas, United Nations demographers estimate that by the year 2050 the world population could exceed 9 billion and perhaps approach 10 billion.[14]

But that's not the worst of it, because in 2014 the United Nations revised its projections of world population growth even higher. According to a report in *Science* magazine, the numbers would reach 9.6 billion by 2050 and 10.9 billion by 2100. The authors gave an 80 percent likelihood that 2100 population would stand between 9.6 and 12.3 billion. Previously, the UN had believed population would peak at around 9 billion and then decline.

That explosion of human numbers is astonishing. Well over a thousand times more people are living today than at the beginning of the Agricultural Era 10,000 years ago (6.7 billion vs. 5 million as of 2010). What a change that is from the pre-farming era during which, as we saw, population may have grown by a mere thirty or so persons per year.

In sheer numbers the rate of growth continues at a dizzying pace. According to recent estimates[15] in 2015 the world population was growing by about 220,000 people every day. That added a number equal to the population of Scottsdale, AZ, each 24 hours.

Every week more than 1.5 million are added to the total, equal to about the present population of Philadelphia, Pennsylvania.

---

[13] Source: United Nations Population Division, Department of Economic and Social Affairs.

[14] *"World Population to Exceed 9 Billion by 2050,"* news release, United Nations Population Division, 11 March, 2009.

[15] *CIA World Factbook,* www.cia.gov.

And every 30 days the number grows by 6.6 million, equal to another Los Angeles *plus* another Chicago.

Yes, as of 2015 each month the world population was increasing by an amount equal to the combined populations of America's second and third largest cities, more than the total world population at the dawn of agriculture. It was adding more than 80 million to the human rolls during that single year.[16]

During those same periods of time—whether days, weeks, months or years—the resources of the Earth did not increase and in fact were steadily reduced by mining and drilling, construction of buildings and roads, logging, slash-and-burn clearance of land, destruction of wetlands, spreading desertification, and the depredations of wind and water erosion.

Clearly, runaway population growth—and can we call it anything else?—is an unwelcome effect of agriculture. With the industrialization of food production, particularly since about the middle of the 20th Century, the explosion of human numbers has metastasized, posing a threat to the future of civilization and the very survival of humanity.

The Misunderstood Prophet

Any discussion of population and environmental issues must begin with Thomas Malthus (1766-1834), the Englishman whose "Essay on Population" was published in 1798. Malthus has often been presented as a radical and perhaps unhinged doomsayer who predicted that the human race would continue to multiply until the Earth was overwhelmed by the press of human numbers. For generations he has been ridiculed for these (supposed) ideas.

---

[16] It is important to keep in mind that these are net figures, births minus deaths, so it means 80 million more hungry mouths for the Earth to support than one year earlier.

Until recently it had not occurred to me that I didn't know anyone who had actually read Malthus, nor had I seen any in-depth discussion of his work. I was particularly alerted when I saw a dismissive book passage describing him as "an English monk." I knew that was wrong and the statement seemed to reflect a good deal of ignorance on the part of the writer.

My curiosity aroused, I determined to find out what this supposed doomsday character actually said. I ordered a copy of Malthus's book.[17] (Yes, book, for we have been misled by the word "essay" in the title and in fact, with added biographical notes what I received is actually a volume of more than 170 pages.) Reading it was a revelation in more ways than one.

First, I learned that Malthus was far more than just an ordinary country parson as he is often portrayed (much less a monk). He was a scholar of some note, having graduated from Cambridge University with honors in mathematics in 1788 and completing an MA degree there in 1791. He was elected a Fellow of Jesus College, Cambridge, in 1793.

In his early years and at the time he wrote the first edition of his famous essay he did indeed earn his living as an Anglican parson. However, he went on to become one of the most influential economists and political scientists of his era. In 1805 he was appointed Professor of History and Political Economy at England's East India College, a position he held until his death in 1834.

Malthus was well traveled, including an extended trip to Scandinavia and Russia in 1799 and visits to Switzerland and France in 1802. He was a founding member of the Political Economy Club (London, 1821) and was elected a member of the Royal Society, England's preeminent scientific institution. He pub-

---

[17] T. R. Malthus, *An Essay on the Principal of Population*, Oxford World's Classics edition, Oxford University Press, rev. 2004.

lished many books and articles on economics and was recognized as one of the leading thinkers of his generation.

Malthus left many marks on his age. It was to his work that Thomas Carlyle referred when he branded economics as "the dismal science." Malthus's correspondence with David Ricardo is recorded as one of the most important controversies in the history of economics. His work also had a significant impact on science by inspiring the work of Charles Darwin, as reflected in this excerpt from Darwin's autobiography:

> In October, 1838, that is, fifteen months after I had begun my systematic enquiry, I happened to read for amusement Malthus on Population, and being well prepared to appreciate the struggle for which everywhere goes on ... it at once struck me that under these circumstances favorable variations would tend to be preserved, and unfavorable ones to be destroyed. The result of this would be the formation of a new species. Here, then, I had at last got a theory by which to work.[18]

So, we have learned that Malthus was far more than just a simple country parson, a characterization that I suspect has been put forth as a way of marginalizing his ideas by those who seek to minimize the threat of over-population.

But what did Malthus actually say about this subject? The answer is surprising, and a review of his ideas is helpful in understanding the principles of population. Initially, Malthus set down two postulates:

• First, that food is necessary to the existence of Man.
• Secondly, that the passion between the sexes is necessary, and will remain nearly in its present state.

---

[18] Charles Darwin, *The Autobiography of Charles Darwin: 1809-1882*, pb. edition, W. W. Norton & Co., 1993, pgs. 98-99.

Next he recognized the mathematical fact that human pop‑
ulation and the means of food production increase by different
progressions, the first geometrically and the latter only arith‑
metically. From that he derived that "the power of population is
indefinitely greater than the power in the earth to produce
subsistence for man." He continued (emphasis added):

> Population, when unchecked, increases in a geometrical
> ratio. Subsistence increases only in an arithmetical ratio.
> A slight acquaintance with numbers will show the im‑
> mensity of the first power in comparison of the second.
>
> By that law of our nature which makes food necessary to
> the life of man, *the effects of these two unequal powers must be
> kept equal.*

From this it is clear that Malthus believed that there was a
natural balance between population numbers and what the
environment can support. To illustrate this point he devised a
thought experiment based on the assumption that the population
of the world were to double each 25 years, and that food pro‑
duction would also increase at its own natural rate at 25 year
increments.

Noting that arithmetical increases in food would follow the
series 1, 2, 3, 4, 5, 6…, while geometrical increases in population
would follow the series 1, 2, 4, 8, 16, 32…, he concluded:

> In two centuries and a quarter, the population would be
> to the means of subsistence as 512 to 10; in three
> centuries as 4096 to 13; and in two thousand years the
> difference would be almost incalculable, even though the
> produce in that time would have increased to an
> immense extent.

So there is Malthus's "doomsday scenario," and it was presen‑
ted only as a thought experiment for the purpose of demonstrating

the *impossibility* of such an event! In fact, Malthus devoted several chapters to showing how he believed population and the food supply would remain in balance.

To briefly summarize, he postulated two kinds of natural force acting to control population: First a positive force that he labeled "misery," including hunger, disease and war, which acted to increase death rates; and second a preventative force, which he termed "vice," in which he included abortion, birth control, and prostitution, all of which tended to hold down the birth rates.

To add dimension to this brief examination of Malthus's thought, here is a pertinent passage from the Essay (emphasis added):

> Through the animal and vegetable kingdoms, nature has scattered the seeds of life abroad with the most profuse and liberal hand. She has been comparatively sparing in the room and nourishment necessary to rear them. The germs of existence contained in this spot of earth, with ample food and ample room to expand in, would fill millions of worlds in the course of a few thousand years. *Necessity, that imperious all pervading law of nature, restrains them within the prescribed bounds.* The race of plants and the race of animals shrink under this great restrictive law. And the race of man cannot, by any efforts of reason, escape from it.

Malthus was an economist and the *Essay on Population* is essentially an economic treatise. He devotes several chapters to the economic limits to population growth, e.g., that when there is a surplus of labor, wages will decline, and that when there is a shortage of produce, prices will rise.

Thus he concludes that the poor will bear the brunt of population pressure, being caught in a bind of "misery" when squeezed between low earnings and rising costs of living. We see this factor at work today among the "beneficiaries" of industrial agriculture.

More than two centuries after Malthus wrote his Essay on Population, we can view his ideas from a perspective that he himself could not. We can see something quite extraordinary — that while he has been unjustly branded as a doomsayer, the very population catastrophe that he thought could not happen may actually be taking place today.

What is the reason for this ironic twist? It's because Malthus could no more have foreseen the rise of industrial agriculture than he could have envisioned a nuclear powered aircraft carrier or the International Space Station. In his time the Industrial Revolution was just beginning to get its boots on. The application of industrial-scale machinery, the use of artificial fertilizer and chemicals, plant and animal genetics, all the tools of the so-called "Green Revolution" worked to (temporarily) undermine Malthus's conviction that misery and vice could act naturally to contain population. Without being able to foretell the future of industrialized agriculture it would have been impossible for Malthus to conceive of the Earth as we know it.

It is interesting to compare Malthus's ideas with modern day warnings of the dangers of over-population that are raised by environmental scientists. Those who wish to deny the dangers posed to our civilization by the collision between population and the environment often paint these modern-day prophets with the same brush as Malthus, but they actually come from different points of view. The ecological scientists of our time are aware of the changes brought by technological agriculture. They recognize that the natural limits envisioned by Malthus have been breached by the industrialization of food production.

Malthus did not know that when humans had begun to

assume the role of masters of the environment 10,000 years ago, rather than living in balance with nature, the link with nature had been broken. In his time two centuries ago the effects were still relatively slight. Today, industrial farming has kicked the disconnection between humans and nature wide open.

In failing to take into account the unanticipated effects of industrial agriculture, Malthus's thought experiment may actually have touched on a version of the real future. A doomsday scenario that he thought could never happen may be coming to pass after all.

Paul Ehrlich's "Population Bomb"

In more recent times and particularly since the end of World War II, a few other brave souls have sounded the trumpets of warning. These include[19] Fairfield Osborn, Harrison Brown, Georg Borgstrom, Lester Brown, Joel E. Cohen, and perhaps most notoriously Paul Ehrlich, whose book *The Population Bomb*[20] first appeared in 1968.

Unfortunately, Ehrlich's message was ill received by many, especially by members of special interests such as politicians, financiers, commodity brokers and oilmen who saw his warnings as a threat to their comfortable positions of status quo. Ehrlich based his scenarios on direct extrapolations of trends of the times, making specific and often doom-laden predictions of looming near-term famine and other disasters.

Although he was well aware of the unfolding Green Revolution and other technological initiatives, he failed to fully anticipate

---

[19] See the Bibliography for references to these and other writers.

[20] Paul R. Ehrlich, *"The Population Bomb,"* rev. edition 1971, Buccaneer Books. An entomologist by training, Ehrlich is Bing Professor of Population Studies, Department of Biological Sciences, Stanford University.

the greater-than-expected success of those actions. When Ehrlich's target dates came and went without the disastrous famines and plagues he predicted, critics attacked his entire thesis and attempted to make the case that the predictions were baseless and would never happen.

As anyone must be aware who has stayed abreast of the subject and has kept an open mind, for the most part it was not that Ehrlich's predictions were incorrect per se (although events have taken some turns he did not anticipate[21]), but merely that his projected timing proved to be off. Unfortunately, the population threats he described are looming even larger today and humankind has gone more than four decades further down the dead end path.

Ehrlich laid the groundwork of his argument by making the point that population expansion is determined both by birth rates and death rates, and that while birth rates have been on the decline, so have death rates. Ehrlich was certainly not the first to recognize this. Due to public health measures, vaccination, and disease control and eradication programs, people are living far longer, especially in the poorer parts of the world. He called this "exported death control," since the medical technologies that have extended life expectancy come almost entirely from the developed world and have had the greatest impact in the so-called developing world.

Ehrlich noted that each time the population doubles the entire infrastructure of food production, transportation, housing and everything else humans require must also be doubled just to maintain the same standard of living as before.

---

[21] For example he foresaw photochemical smog as a growing problem and predicted oxygen stations would appear on street corners in Los Angeles. The Clean Air Acts of 1963, 1970 and 1993 corrected that problem and collectively serve as an example of how environmental problems can be solved.

That's bad enough but it gets worse, for to provide the increased standards of living many people have come to expect, the infrastructure must actually be tripled to support a doubling of population.

Again in line with Malthusian thought, it is a familiar argument that such a scenario cannot repeat itself over and over again. There is only so much arable land, only so many natural resources, only so much room on the Earth.

Ehrlich concluded: "No growth rate can be sustained in the long run." This points to static or declining populations as the only viable options.

He poses that there are only two possible "solutions" to the challenge of bringing population under control.

One, the birth rate solution, could have been developed through deliberate action. Although never organized as a global plan, that option has naturally emerged to some extent — particularly in the advanced nations and through China's one-child policy — but to date the decreases in birth rates have been insufficient to halt, much less reverse, the growth in human numbers. In most of the world the population bomb is still ticking.

In the alternative "death rate solution," Nature finds its own ways to reduce population through war, famine and disease. Here Ehrlich recapitulates Malthus's concept of "misery" as a controlling factor to population.

Should humankind ultimately fail to act sufficiently on the "birth rate" option, nature is steadily and surely working her way toward the "death option".

In retrospect, this point of Ehrlich's seems entirely relevant today in a world beleaguered with war, resource depletion, climate change, and spreading hunger. By 2008, a headline in the news-paper *USA Today* was sounding the alarm: "Surplus U.S. food supplies dry up."[22]

Now let's examine Ehrlich's direct predictions. As was the case with Malthus, what he actually said is at odds with what his critics would have us believe. First and foremost is his chapter titled "Too Little Food," in which he stated: [S]ometime around 1958 the stork passed the plow."

In other words, in light of what was known in the late 1960s, he believed that the world had more people than the planet was capable of feeding on a sustained basis. The Green Revolution burst of food production was already beginning, but in Ehrlich's opinion at that time, it could "at the very best buy us only a decade or two in which to stop population growth." He pointed out that in 1969, despite Green Revolution measures, there was no increase in the world's food production while population rose by two percent.

Next Ehrlich stated:

> "[I]f the pessimists are correct, massive famines will occur soon, possibly in the 1970s, certainly by the early 1980s. So far most of the evidence seems to be on the side of the pessimists, and we should plan on the assumption that they are correct."

Famine did not come as predicted, and in fact food surpluses were a problem for American industrial agriculture right through the 1980s.

It's interesting to note that he did not make these predictions directly, as has generally been said of him, but merely pointed to the fact that he shared the pessimistic views of others and that there

---

[22] USA Today web edition, May 2, 2008.

was evidence to support those views. Thus, Ehrlich was not a lone voice in the wilderness at all, but was expressing an established, albeit perhaps minority point of view.

We have seen that those views proved incorrect in regard to timing. That was because he underestimated the power of technology, at least for the short term. The Green Revolution launched in the 1940s was only the start of a massive ramping up of food production through the use of plant breeding, pest control, increased use of artificial fertilizer, genetic engineering, and the opening of millions of acres of new lands to the plow. In the end, however, we have to conclude that Ehrlich's basic assumptions were valid and that the predicted events failed to occur only because the estimated time frame was too short.

Adjusting Ehrlich's Predictions

As an intellectual exercise, let's adjust those figures. Instead of assuming as Ehrlich did that the application of industrial agriculture would delay famine by only one to two decades, let's expand his estimate by a factor of three, placing the projected time of trouble at 30 to 60 years after his book appeared. That points to the period from 1999 to 2029—the very time in which we are living.

In light of present knowledge and on-going trends, in this adjusted time frame his predictions seem far more likely. In fact, more than one billion people—as many as existed on the entire Earth less than two centuries ago—already are malnourished and living on the edge of famine.[23] Each year many die of starvation and diseases resulting from poor nutrition as the world's food supply draws tighter. In many poorer nations life expectancy is beginning to fall once again.

---

[23] "More people than ever are victims of hunger," op cit.

The huge crop surpluses of the late 20th Century are a distant memory and it may not be long before the few remaining exporters of food, mainly the U.S., Canada and a handful of others, will be able to do any more than feed their own people.[24] In 2008 a worldwide crisis erupted as food prices soared due to shortages, surely an ominous portent of future events.

Twenty-two years after the publication of *The Population Bomb*, Ehrlich along with his wife Anne Ehrlich, returned to the subject of population.[25] Emphasizing the speed with which exponentially-driven events can sneak up on us,[26] they wrote: "The key point to remember is that a long history of exponential growth in no way implies a long future of exponential growth."

The Ehrlichs pointed out in 1990 that even though the fertility rate of the human race had showed some signs of decline during the 1980s, the projected doubling time of the world population had only risen from 33 years to 39 years. They continued:

> "So even though birthrates have declined somewhat, Homo sapiens is a long way from ending its population explosion or avoiding its consequences. In fact, the biggest jump, from 5 to 10 billion in well under a century, is still ahead. But this does not mean that growth could not be ended sooner, with a much smaller

---

[24] The ominous practice of converting food crops into ethanol or biodiesel has put the approaching possibility of widespread famine onto a fast track.

[25] Paul R. Ehrlich and Anne H. Ehrlich, *"The Population Explosion,"* Simon and Schuster, 1990.

[26] This is the same effect to which Malthus pointed, the difference between geometric or exponential increases as opposed to arithmetic ones. A simple graph of an exponential process against time shows a line that rises slowly at first before soaring upward like a rocket. To think of it numerically, doubling a small beginning figure seems to yield a relatively insignificant amount (e.g., 1 + 1 = 2), but as the base grows larger, doubling a very large number further on during the same span of time yields a significant increase (e.g., 3 billion + 3 billion = 6 billion).

David L. Brown

population size, if we—all the world's nations—made up our minds to do it. The trouble is, many of the world's leaders and perhaps most of the world's people still don't believe that there are compelling reasons to do so. They are even less aware that if humanity fails to act, nature may end the population explosion for us—in very un–pleasant ways—well before 10 billion is reached."

Denial of the Population Problem

This highlights one of the most troubling things about the present world situation: the widespread denial that over-population threatens the environment of our planet and the very future of humanity. What is it that people don't understand about simple statements of obvious fact such as: "There are limits to how many people the Earth can support"?

Every day we read or hear examples of a wounded and over-exploited environment, strong evidence that Nature is in retreat before a growing crowd of humanity. Ocean fisheries are collapsing from aggressive harvest. Rain forests are falling to the roar of chain saws. Deserts are spreading and the seas are rising. Oil and other energy sources are peaking and becoming scarce and more expensive.

All that and much more goes back to over–population and the exploitation of resources that keeps our numbers growing.

Despite these warning signals, there are many who decry the fact that population growth is slowing, particularly in advanced societies. As an example of this point of view, which reflects a belief that falling fertility rates are a dire threat to humanity, here are the words of David Reher, a population historian at the Complutense University of Madrid, Spain:

> We are in the midst of a cascading fertility decline. Even a TFR [total fertility rate] of 1.7 is not safe; it is a disaster

38

if you look a couple of generations down the line. …
Urban areas in regions like Europe could well be filled
with empty buildings and crumbling infrastructures as
population and tax revenues decline. … [Due to aging
and labor shortages] it is not difficult to imagine enclaves
of rich, fiercely guarded pockets of well-being surrounded by large areas which look more like what we
might see in some science-fiction movies.[27]

That's a dire prediction indeed—but is it reasonable? I could
raise a number of questions here, including the observation that
immigrants from Africa and Asia are already filling all those
European buildings and will no doubt continue to do so. But my
essential argument against such positions is that they are based on
the assumption that healthy economies can exist only in conditions
of GDP growth, and that ever-increasing populations are required
to fuel economic expansion. Can that be true that the only sound
economic policies are those that grow populations beyond the
ability of the Earth to sustain them? That does not seem logical,
but I'll leave the question for now. I discuss this and other economic issues in later chapters.

Growing Numbers Mostly in Poor Regions

While population growth rates have begun to decline, and
even reverse in some advanced nations such as Japan and Italy, the
overall picture remains grim due to continuing high birth rates in
the less developed parts of the world. In 2005, for example, there
were some 137 million babies born in the world. Of those, only 13
million were born in the entire "rich" world, including Europe,
Japan, the United States, Canada and Australia. That number
includes children born to recent immigrants to those places.

The other 124 million children were born in "emerging" or

---

[27] *"The Baby Deficit,"* Science, 30 June, 2006, pgs. 1894-1897

"developing" countries such as China, which alone produced 16 million. More than 90 percent of all the newborn babies on Planet Earth arrived in poor or so-called developing nations. Indeed, the ten most populous countries in the world include China, India, Indonesia, Pakistan, Bangladesh and Nigeria.

Isn't there something alarming about the fact that by far the largest population growth is taking place in those regions of the world that are least capable of supporting the new babies? What kind of future can they have as resources continue to shrink in the face of growing masses of people and declining resources of food?

China deserves special mention here, not only for its position as the world's most populous nation but also for its attempts to rein in population growth. China claims that its family planning efforts have "curbed fast population growth and prevented 400 million births." According to official pronouncements, this would have postponed the nation's "drive to build a well-off society."[28]

However, in view of China's 2015 population in excess of 1.4 billion, it is interesting to read the following statement written in 1988 by Jian Song and Jingyuan Yu, two professors in Beijing who studied China's projected carrying capacity since 1950. They wrote: "…the long-term strategic goal of China's population policy should be the control of the population at a figure below one billion, or ideally, below 700 million."[29]

At the time they wrote, the population of China had already exceeded one billion. The barn door was standing wide open and the horse was nowhere to be seen.

---

[28] Comments attributed to Zhang Weiqing, director of the National Population and Family Planning Commission, *Xinhua News*, 3 May, 2006.

[29] Quoted by Joel E. Cohen, *"How Many People Can the Earth Support?"* W. W. Norton & Co., 1995.

In view of continuing population growth in poor or so-called developing nations it is also worth noting that the relatively few advanced countries are favored destinations for "economic immigrants" striving to escape from poverty or famine. This is particularly true in Western Europe and the United States and Canada. As famine and environmental problems spread through the undeveloped parts of the world, the pressure to migrate will become ever greater. It is an unfolding problem.

Signals such as these are many and ominous, and yet most people in the advanced parts of the world choose to pretend that everything is just fine. It's business as usual. Even though we passengers on Planet Earth are metaphorically riding the Titanic, many fail to recognize the dangers and refuse to accept that over-population caused by industrial agricultural lies at the heart of the problem.

What Are the Limits of Population?

Despite that evidence, there are astonishingly diverse opinions on the subject of how many humans the planet can support, ranging even as high as one trillion and, in at least two cases, even far higher. It's hard to see how these numbers could be derived, but remember the words of Malthus who said that unlimited expansion of life, if it were possible, would fill millions of worlds in only a few thousand years. Perhaps the wildly optimistic pre-dictions are based on the potential of our race to multiply, without considering the physical limits on that growth.

It's instructive, and even amusing, to see just how far some people have gone to paint a rosy picture. A population of one trillion, for example, would be more than 140 times larger than at present and would represent a world beyond our recognition. Multiplied by that factor, the population of the United States would be more than 40 billion. One shudders to think of the conditions in places such as India and China, much less desert Egypt, which in 2015 had more than 87 million people on a land that is less than three percent arable and with only about a sixth of that in permanent crops.[30]

Incredibly, there have been even more optimistic (if we can call them that) estimates, including one made by physicist J. H. Fremlin of the University of Birmingham, England, who reportedly set 10^{16} to 10^{18} as the range of possible human population.[31] Those are staggering figures. Stated in plain numbers they are equal to 10 quadrillion to one quintillion.

A quick calculation shows that even the lower number would place more than 170 million people on each square mile of land on Earth, including deserts, mountains, and even Greenland and Antarctica.[32]

Hard though it may be to believe, an even more bizarre statement on this subject was made by the late Julian Simon, a former professor of business administration who was a senior fellow at the Cato Institute, a libertarian think tank. He wrote extensively on economics, population and resources and is often quoted by free market Libertarians due to his dismissive views about the limits of the Earth.

On population, he wrote: "We now have in our hands—really,

---

[30] CIA Factbook, op. cit.

[31] Joel E. Cohen, op. cit.

[32] A square mile contains nearly 28 million square feet, so this would pack about six people on each square foot of the total land area of the Earth, allowing about 24 square inches for each person. That would allow no room for anything else, including farmland, houses, schools, shopping centers, roads, tennis courts or parking places. And that is Prof. Fremlin's *low* estimate! At the high end of his range, things would start to get seriously crowded with 600 people per square foot. At that rate, my house (not counting the yard or garage) would be home to more than a million people.

in our libraries—the technology to feed, clothe, and supply energy to an ever-growing population for the next seven billion years."[33]

My electronic abacus doesn't have nearly enough beads to calculate the immense figure that would result from the continued compound growth of human numbers for seven billion years.[34] Such a possibility is beyond ridiculous and has absolutely no basis in reality.

On a more reasonable level I have seen estimates that even in its original state the Earth could have supported no more than about two billion humans for any extended period, and certainly not for centuries or millennia to come. I suggest that in the longer term there is good reason to question even that number as optimistic, perhaps vastly so. As we have seen, in the actuality of humanity's original state prior to the coming of agriculture, our planet supported around five million people.

Obviously, the Earth is no longer in its original state because it has been sorely wounded and degraded by human action. Two billion is 400 times the estimated 5 million humans present at the beginning of agriculture, living as hunter-gatherers.

Future technologies and lifestyles may set the eventual number higher or lower, but in the end it doesn't really matter how many people will share the Earth, only that their numbers must be and remain in balance with the resources available to them.

---

[33] Julian L. Simon, *"The State of Humanity: Steadily Improving,"* Cato Policy Report, Cato Institute, September/October 1995.

[34] I challenged an atomic scientist friend of mine with this and even his advanced scientific calculators could not handle such an enormous number. One topped out at 10^{200} and his "big gun" went as high as 10^{500} without being able to estimate the number of people that might be alive after seven billion years of continued population growth. For comparison, there are around 4×10^{80} atoms in the Universe. Such is the power of compounding.

David L. Brown

The Tragedy of the Commons

One of the most influential essays ever published in *Science*, the journal of the American Association for the Advancement of Science, was titled "The Tragedy of the Commons,"[35] by noted ecologist Garrett Hardin. I say "influential," but like so many seminal ideas it actually influenced a small proportion of the total population of the planet, mainly scientists and a few enlightened intellectuals.

In the essay Hardin argued that the Earth is not without limits, that "Space is no escape," and that "A finite world can support only a finite population: therefore, population growth must eventually equal zero."

To illustrate his point he used the metaphor of an English "commons" to represent the Earth as a whole. Commons were areas of land open to all and on which every herdsman could place more cattle or sheep until the carrying capacity of the field was exceeded, whereupon all were ruined.[36]

As Hardin recognized, humanity has gone from an existence comfortably in harmony with nature to the unsustainable absolute domination of the ecology of the planet. To support ourselves and our domesticated animals we have put vast millions of acres of prairie to the plow, drained countless wetlands, paved over entire landscapes, put millions more acres of forest to the chainsaw and axe, and are despoiling the atmosphere and oceans with greenhouse gas and toxins.

---

[35] Garrett Hardin, *"The Tragedy of the Commons,"* Science Magazine, 13 December, 1969, pp, 1243-1248.

[36] In England the problem eventually was solved through the process of "enclosure," in which lands were fenced and deeded to individuals. Many small-scale farmers were forced to leave the country and became grist for the emerging factories of the Industrial Revolution.

Overshoot-and-Collapse Could Happen To Us

There is another population factor at work here, something
ecologists call "overshoot-and-collapse." This describes a phenomenon that occurs when a group of animals in a restricted
environment becomes capable of breeding without limit. When
food is temporarily abundant and there are no predators or other
factors to hold it in check, numbers zoom past the ability of the
environment to support them. That leads to the ultimate collapse
in the form of a rapid and disastrous die-off.

Here is an account of the classic example of overshoot-and-
collapse:[37]

In 1944, toward the end of World War II, the U.S. Coast
Guard was maintaining a 19-man outpost on St. Matthew Island in
the Bering Sea. Noticing that the island had a rich cover of lichens,
a favored food of reindeer, and desiring to provide a backup food
supply for the men in the outpost, the Coast Guard brought 29
reindeer to the remote island and released them to feed on the
lichens.

The following year the outpost was abandoned and the
reindeer were left to their own devices until 1957 when U.S. Fish
and Wildlife Service biologist David R. Kline visited the island. He
was no doubt astonished to find a thriving herd of 1,350 reindeer
feeding on a four-inch thick mat of lichens that covered the 128-
square-mile island. There were no wolves or other predators to
restrain the growth of the reindeer population and, at least for the
time being, plenty of food to sustain them. For these animals, the
"population bomb" was ticking loudly.

---

[37] David R. Kline, *"The Introduction, Increase, and Crash of Reindeer on St.
Matthew Island,"* undated paper, Alaska Cooperative Wildlife Research Unit,
University of Alaska. The paper covers Kline's observations through 1966.

When Kline again visited the island six years later he observed that the number of reindeer had exploded to around 6,000, an unsustainable deer population far in excess of what the lichen crop could support. "Overshoot" had taken place, and "collapse" was now inevitable.

Indeed, only three years later when Kline paid another visit in 1966, he found the island strewn with the skeletons of dead reindeer and hardly any lichen remaining. He counted just 42 surviving reindeer, 41 females and one sickly male. There were no fawns. By sometime soon thereafter the last reindeer was gone.

Overshoot-and-collapse sends a strong message about how important it is for populations—whether of deer, rabbits, wolves or human beings—to exist in a sustainable balance with Nature. Whenever that balance is disturbed, trouble is sure to follow.

Human civilizations have not been immune to the destructive effects of overshoot-and-collapse. In a recent book, geographer Jared Diamond described the collapse of individual societies when they over-ran the ability of their piece of the Earth to support them.[38]

A Pulitzer Prize winner and professor at the University of California-Los Angeles, Diamond documented the failure of societies as diverse as those of the Anasazi of the American Southwest, the Easter Islanders, the Maya and Viking settlements in the New World.

He asked the question: "Why do some societies make disastrous decisions?"

Well, the answer is not simple. For one thing, no decisions

---

[38] Jared Diamond, *"Collapse: How Societies Choose to Fail or Succeed,"* Viking Penguin, 2005.

grounded in an understanding of overshoot-and-collapse were involved, any more than the reindeer of St. Matthew Island chose to breed themselves into a dead end. Whether humans or deer, these prior victims of overshoot-and-collapse were just doing what came naturally, ignorant of the possible consequences.

We can see that overshoot-and-collapse can affect entire human societies on a limited scale, such as an island, an isolated region or a nation-state. But what if the entire Earth is becoming the stage for the greatest overshoot-and-collapse in history?

That question deserves our considerable attention, for as we have seen there is ample evidence that the number of human beings that the Earth is capable of supporting has already been passed.[39]

If we can envision the entire planet as having a limited carrying capacity (which is definitely true), and humanity as having exceeded that capacity (which we probably have) and with nowhere else to go (there is neither more land to be discovered nor do we have a future in space), then it is more than just a metaphor to compare our situation with that of the reindeer population on St. Matthew Island. It appears more than possible that overshoot-and-collapse provides an actual working model for what humanity faces today.

There is a difference, though, for we are not helpless animals living by instinct and chance alone. Yes, the collapse of earlier human societies such as Diamond described is cause for concern, but we are a clever species with a vast store of common knowledge.

---

[39] For example, *Global Footprint Network* (www.footprintnetwork.org) estimates that we humans already are consuming the product of about 1.3 planet Earths each year. Their website states that "Moderate UN scenarios suggest that if current population and consumption trends continue, by the mid 2030s we will need the equivalent of two Earths to support us. And of course, we only have one."

Since those historical collapses took place we have made vast strides in our understanding of Nature, so there could be hope for us yet. The challenge is to apply all our accumulated wisdom and knowledge to work our way through the possible collapse that could follow the present state of population overshoot.

So far, unfortunately, the picture is disheartening. Science and technology, particularly as applied to the industrialization of agriculture, have actually contributed significantly to the problem of population growth. Can our clever skills and tools be turned to a different set of goals and put to work to help soften or forestall the impending collapse? Some seem to think so, but time is running out.

If it is true that humans have already exceeded the ability of Nature to support our species, the urgency of the situation cannot be over-stated. An even more dire fact is that we are rapidly running down the remaining resources of the planet to support continued, unsustainable population growth.

The Importance of Resources

As non-renewables continue to be used up—an inexorable process that's taking place right now—industrial agriculture, and indeed our civilization as a whole, must soon begin to move toward greater sustainability. Eventually, as resources are completely depleted or become too difficult to extract and use, common sense tells us that the model must eventually become completely sustainable, using only renewable resources or those that are present in essentially infinite supply (e.g., stone, wood, plant fibers, animal products, clay, sand, sunlight, water and wind). Even fertile topsoil cannot be considered a renewable resource, since it takes many hundreds of years for nature to create even an inch of it.

In the decades ahead and henceforth to the end of our time on

Earth, absent completely unexpected technological breakthroughs
our descendants will not be able to rely on petroleum, natural gas,
coal, iron, phosphate, copper, zinc and all other non-renewable
resources now considered essential.

Many if not most economists consider these concerns to be
trivial, since their theories presume that an equally suitable
replacement will always be found for any depleted resource. Their
presumptions will be examined later in this book, but it seems
unlikely that all the "essential" non-renewables on which our
civilization depends will be miraculously replaced by hitherto
unknown substitutes that are at least as good.

If that is true, then the inevitable depletion of the natural
resources on which industrial agriculture is based will demand that
we must begin to wean ourselves away from our present approach
to food production—and with Peak Oil perhaps already upon us,
that process must begin now. We'll examine the critical subject of
resource depletion in the next chapter.

David L. Brown

Chapter Three

An Empty Cornucopia

Virtually every tangible thing we own, use, or consume in the course of everyday life is the product of industrial technology and has been created through the destruction of natural resources. That applies throughout our civilization, from the clothing we wear, the roofs over our heads, the automobiles we drive, and the computers we use to do our modern version of "work." It applies to the most mundane things: a paper clip, a coffee cup, a home-cooked meal, even this book.

Try to think of anything that comes into your life directly from the natural world, rather than through the processes of industrial technology. Chances are you'll be hard pressed to come up with much.

If you're a gardener, thinking of your ripe strawberries and tomatoes, you may think, "Aha! I've got him there." But think again, because your fruits and vegetables most likely rely on the use of plant fertilizer, insecticide, herbicide, and a basket-full of other products of the industrial world. And even if you're the most avid of organic growers you most likely use rakes, hoes, dibbles, hoses, sprinklers, buckets, shovels and wheelbarrows — all manufactured products. I rest my case.

Industrial production is based on the "development"[40] of natural resources, which is to say those resources are exploited and turned into products for the consumption of human beings. That defines the essence of the dead end path, for the world has only a limited supply of resources. We cannot continue forever to plunder the irreplaceable riches of our planet. If virtually all the things we own and use are produced through exploitation of resources, then to understand the situation we face we must focus our attention there.

I chose the metaphor of an empty cornucopia for resource depletion because it reflects the grim possibilities of the future. The myth of the cornucopia, the "horn of plenty," arose in fifth century B.C. Greece as a symbol of food and abundance, assurance from the deity Zeus that nature would always provide. No matter how much was used, there would always be more.

And so it seemed in those simpler times, but that was then and now we are more than two centuries into what has been called the Industrial Age, and we are ten thousand years down the dead end path of agriculture, an awakened monster that has run amok.

Clever humans have devised ways to take the resources of nature and use them to make the "things" and "stuff" that fill our lives. It is the rare individual today who lives "close to Nature." Few of us have even been in the neighborhood and sometimes it seems that we must be on some entirely different planet, but sadly it is poor old Earth that we inhabit.

_____

[40] Like "progress" and "improvement," "development," is one of the code words through which destruction of the natural world is presented in a favorable light. In this vein the late Edward Abbey, author of "The Monkey Wrench Gang" and a life-long advocate for the environment, called the U.S. Bureau of Reclamation, builders of great dams and other projects that scar the face of the Earth, the "Bureau of Wreck the Nation."

Those who "drop out" of mainstream life to become organic subsistence farmers or modern day pioneers are thin on the ground. The few remaining hunter-gatherers of the Earth, the last humans who have some idea of how to live in harmony with nature, are scorned as primitive savages. The last of them are rapidly being drawn into lifestyles whose boundaries are defined by technological civilization —or suffer far worse fates.[41]

Is Progress Unreasonable?

The writer and dramatist George Bernard Shaw once wrote: "The reasonable man adapts himself to the world; the unreasonable man persists in trying to adapt the world to himself. Therefore, all progress depends upon the unreasonable man."[42]

Today Shaw's witty statement reflects a paradoxical view, but in his time "progress" was deemed to be good, a positive force. Shaw probably didn't recognize the fatal flaw in the concept of "progress," a model that sets perpetual growth and expanding GDPs as the true measures of successful societies.[43] Of course, as we are coming to realize, this flies in the face of the fact that economic growth is by its nature unsustainable due to the fact that resources are not in infinite supply.

The dictionary defines "unreasonable" as follows:

1 a : not governed by or acting according to reason. b : not conformable to reason : absurd. 2 : exceeding the bounds of reason or moderation.

---

[41] "Pygmies beg UN for aid to save them from Congo cannibals," *The (London) Times*, online edition, May 23, 2003.

[42] G. B. Shaw, "Man and Superman," a play, 1903.

[43] Many if not most economists today still hew to this line, as I will address in a later chapter. Note that "progress" is another code word for destruction of resources.

If Shaw was right that unreasonable men[44] are at the forefront of progress, and one can make a good argument for it, then we can state that our entire technological and economic civilization is the product of a great number of ill-considered actions spread over a period of centuries since the Renaissance and even far earlier.

Ask yourself this: Would you feel safe driving a car designed and built by unreasonable men? Would you trust unreasonable men and women to plan and govern your nation? Would you feel comfortable eating food processed by unreasonable agri-business companies dedicated to the bottom line through theories devised by unreasonable economists?

Well, trick questions aren't they, because of course by Shaw's maxim everything around us is the product of unreason, absurd–ities exceeding the bounds of moderation.

Of course we accept those products. We seldom if ever stop to think whether it was a good idea to go down the dead end path of industrialization, or to weigh whether our civilization might be the result of unreasonable men and women seeking to mold the world to their own convenience and profit.

Our Unnatural Technological World

I sit in an articulated high-tech chair in front of a 23-inch monitor for a Macintosh computer. I have in my view a second Mac with monitor, a scanner, two laser printers, a modem connecting me to the internet, a router to communicate with my

---

[44] A note for readers who may bristle at the use of words such as "men" and "mankind": Except where obviously intended to identify a member of a particular gender I consider such terms to refer to the human race as a whole. I will strive to use neutral or all-inclusive terms such as "humanity" or "men and women" where appropriate, but may also indulge in shorthand usages as in this case, where I am following Shaw. None of these references are intended to transgress the political correctness surrounding gender.

other computers, a half-dozen hard drives, several telephones including an iPhone, three or four electronic abacuses as I call them, an electric stapler, a digital recorder, a device to sense incoming fax signals on one of my phone lines and automatically shunt them to a fax machine, a clock that automatically sets itself by tuning to a radio station somewhere in Colorado, and a set of speakers with which I can hear a collection of digital music residing on my computer and through which I am presently listening to an album titled "East" by Tangerine Dream.

On the opposite wall stands a bookcase containing nearly 150 volumes of reference for the writing of this book. Atop the bookcase sits a toy gorilla capable of enthusiastically singing the Macarena (although his battery has been turned off for years) and a soft toy representing my spirit guide, Ratbert. On one wall there is a black-and-white print autographed by President Jimmy Carter and picturing him shaking my hand at the White House. On another is a color photo of my wife with the private plane we no longer own but often miss. (Well, *I* miss it.) There is a coffee cup featuring a single happy black sheep amid a flock of white ones.[45]

All these and more are products of industrial technology. With the exception of the black-and-white print, the coffee cup and the bookcase and its contents, none of these are things that my grandfathers would have known, not a single one. Together they demonstrate how we denizens of the 21st century live and work. It's just the way things are—but not perhaps the way they should be.

Can it be true that by creating technological civilization—and in particular the subject of this book, industrial agriculture— unreasonable individuals have led us down a dead end path? I am convinced of it; others may disagree. If you are unsure, let your

---

[45] This description was written for the first edition (2010) and while it no longer applies to my situation I have decided to leave it in place for its illustrative value. For example I now use a 27-inch iMac computer.

mind open to the possibility that I and you and everyone else are caught on this dead end path that leads to … well, what?

If unreason started us down this path, then to understand what has happened we must apply the opposite lens of reason. Only when seen in the stark illumination of sound, rational logic and a strict examination of the facts can the errors and delusions of unreason become clear.

Reason is the only tool through which we can begin to understand the world around us. Writer and philosopher Ayn Rand put it this way: "Man cannot survive except by gaining knowledge, and reason is his only means to gain it. Reason is the faculty that perceives, identifies and integrates the material provided by his senses."[46]

The picture is clouded because those "men of unreason" (and yes, women too) are still among us, spreading denial and hoping that the "good times" will always be with us. It's my conviction that they are doomed to disappointment. Unless you are truly a denier, you must admit that at some point there are limits to what Nature can provide, that the resources of the Earth are less than infinite.

The cornucopia cannot provide forever, and when the gifts of nature run out our comfortable way of life will "come a cropper" as the British would say.

Glimpsing the Future of Resources

Let's take a look at this cornucopia of natural resources and see what remains for us. Will there be plenty for centuries to come, or are we approaching the end?

---

[46] The statement is from Rand's fictional character John Galt in the novel *Atlas Shrugged*, and is quoted in *For the New Intellectual*, Rand's book about her philosophy of Objectivism, Signet Books, 1961.

First principles: There are two kinds of resources—those that are renewed by natural processes, and those that are in fixed supply and once used will be gone forever.

Renewable physical resources include fertile soil and growing things such as grass and trees. They include animals that are the source of food, fur, milk, hides, and companionship, or which can do work such as herding or pulling wagons. We may also include those things that are in such large supply that they will virtually last forever, such as stone, sand, clay, and water.

Almost all renewable energy comes through the rays of the Sun, either directly or indirectly in the forms of plants using photosynthesis, flowing water, waves, and wind. Human labor by the strength of muscles is renewed through successive generations.

Non-renewable resources are all those things that are dug from the Earth and used up—things like petroleum, natural gas, iron, tin, copper, lead, aluminum and a host of other commodities.

And we must add to the list some items that are in principle renewable (and indeed mentioned above as such) but in fact are in danger of being used up in the near term, physical assets such as fertile soil, great forests, healthy grasslands, animal species. If over-exploited these may renew themselves—but only on timescales of hundreds, thousands, or even millions of years.

When you add up the columns on each side of the ledger, it paints a grim picture for our technological civilization. Most of the things that drive our industrialized world are in the "non-renewable" column, and many of those in the "renewable" column are being grossly over-used.

How did we come to the point where almost everything in our civilization depends on the destruction of natural resources? It happened gradually, innocently, with no hint of the eventual trouble to come. Imagine a primitive tribe of humans, living in balance with nature. They learn to make fire. They discover how to harden the points of wooden spears in their campfires. They invent cooking. They learn to chip flint to make weapons and tools. They paint pictures of the animals they hunt on the walls of caves. They form primitive pots and bake them in their hearths. They invent the bow and arrow.

And during the first stirrings of agriculture they make stone axes to clear forests, digging hoes to uproot the soil, flint-edged sickles to cut the ripe grain. All these things are early forms of technology: the first baby steps down the dead end path.

A World Dependent Upon Petroleum

Let's focus our attention on what is presently the most important resource relating to industrial agriculture: Petroleum. For thousands of years, human and animal power were the primary inputs to farming. But that's truly ancient history, right? Not really, for you might be surprised at how recently the use of non-renewable energy has become the absolute norm.

Here's a clue: When I was a child on my paternal grandfather's farm in northern Indiana, one of my earliest memories is of him with his team of draft horses. It was 1944. I was three years old and my grandfather was nearly 70, but still fit. Born in 1875, he knew no way to farm except with a team of horses.

My childhood memory invokes a scene at dusk inside the central aisle of the big old traditional barn. It's chilly and the massive horses are still in their harness. One was named Monkey but I have forgotten the other's name and there are none left alive to fill that gap in my memory. Steam rises from their flanks as my grandfather removes the collars and traces and puts them away from an afternoon's work.

Yes, I saw it with my own eyes: horse farming was being done less than a lifetime ago.

Industrial agriculture was just beginning to gain momentum when my grandfather was born. Just two admittedly long generations separate me from a time when petroleum was used to light oil lamps and men and animals did most of the work on farms in America and around the world. With slow and incremental changes, that was the way farming had been done for millennia.

In 1823 three Russian brothers, Vasily, Gerasim, and Makar Dubinin, learned to purify kerosene from crude oil. In 1854 scientist Benjamin Silliman of Yale University was the first American to "fractionate" petroleum into useful components. In 1859 Edwin Drake drilled a 69-foot deep well near Titusville, Pennsylvania. It yielded 40 barrels of crude per day and the Age of Oil began to emerge.[47]

At first the "Black Gold" was used as a source of kerosene and lamp oil, replacing expensive whale oil to light homes and businesses. Meanwhile, during the last part of the 19th century steam-powered traction and stationary engines, their fireboxes burning wood or coal, began to appear on farms.

Just one more step was required to put industrial agriculture on the fast track, and that was soon to come: The introduction of the first practical internal combustion engines late in the 19th century.

Coal remained the world's foremost fuel even into the 1950s, but oil had found its place as the cheapest, most efficient, most convenient, most powerful source of energy for the development of technological farming. After the turmoil of the Second World War,

---

[47] Source: Wikipedia, "History of Petroleum,"
http://en.wikipedia.org/wiki/History_of_petroleum

industrial agriculture went into high gear and humanity began to hurtle ever faster down the dead end path, riding on a "perfect wave" of oil.

Petroleum fueled the rapid growth of farm efficiency and productivity, creating a social transformation that put hundreds of millions of people around the world out of work on the land and sent them into cities to labor in the factories that were the other end of the transformative industrial world.

The shift of population away from the land was unprecedented. In 1790, 90 percent of the new American nation's population lived and worked on farms. By 1860 the number had plummeted to 58 percent; by 1900 to 38 percent; by 1950 to 12.2 percent.[48]

As if that wasn't enough, in just 38 more years the percentage of Americans among the farm population plummeted to just two percent.[49] Thus did industrial agriculture change the rural landscape and send our civilization speeding down the dead end path, driven by the ever-increasing destruction of resources.

According to those who believe in "progress" as a good thing, this period of history was a triumph of humanity. It certainly seemed so at the time, to most observers at least. But if we remember that resources are not infinite and that petroleum is a nonrenewable resource, we should have seen it coming that we must eventually prepare for an end to this joy ride.

When will that ride begin to reach its end? Opinions vary but if we pay no attention to those in the oil business who want us to

[48] Source: "Growing a Nation—The Story of American Agriculture," on-line teaching resource at http://www.agclassroom.org/gan/timeline/farmers_land.htm

[49] "Farm Dwellers' Population Declines to 2% of U.S. Total.," *The Washington Post*, February 10, 1988.

believe the ride will just go on and on for years to come while they continue to pocket hundreds of billions or even trillions in profits—then the beginning of the end of the Age of Oil is probably already here. Oil production has peaked, certainly in terms of light crude. Producers can no longer hold down prices merely by pumping more oil, and thus the future is certain to be far different from the past.

Quite simply, we will never again see cheap and plentiful oil, and because industrial agriculture depends so heavily on petroleum, cheap and abundant food is also coming to an end. Petroleum and food are joined at the hip, because as we have seen it is on the back of a wave of oil that industrial agriculture has hastened us down the dead end path.

How the Oil Peak Threatens Civilization

We'll take a closer look at energy sources and their roles in food production in a later chapter, but for now it's worth noting that oil is perhaps the most critical non-renewable resource and its looming scarcity is a major threat to civilization as it now exists. The history and projected future of supply-and-demand for petroleum is instructive of resource depletion in general. Let's take a look at how resource peaks work with oil as an example.

Now those who seek to divert our attention from the serious nature of this situation will be quick to tell us that there is still a lot of oil in the ground. They will point out that reaching Peak Oil doesn't mean there's no oil left, only that we have used up perhaps half of what there is. They will also tell you that new technologies in discovery, drilling, and processing will no doubt yield even more oil in the future.

Well, as a skeptical English friend of mine would say, that's as maybe.

David L. Brown

Back in the late 1940s a Shell Oil Co. petroleum geologist named M. King Hubbert predicted that oil production in the U.S. would peak by about 1970.[50] He based that conclusion on his observation that production would follow a bell curve pattern, rising to a peak and declining once more as demand ate up supply. The now-famous Hubbert Curve predicted the actual events with great accuracy. Although his analysis was generally ignored at the time, in hindsight no one can doubt that Hubbert hit the nail squarely on the head.

Now that Peak Oil has become past history in the U.S., what about in the rest of the world? Well, credible evidence suggests that the overall oil peak has already been reached and we are teetering on the edge of the downward slope.

For example, the Energy Watch Group (EWG), an independent organization headquartered in Germany, published their version of an up-to-date Hubbert Curve as of 2008 for the entire world's petroleum production, broken down by regional sources.[51] The graph shows that according to the EWG the world actually passed the Oil Peak sometime around 2006. Production records appear to confirm this. It goes far to explaining the extreme oil price spike during 2008 and the resulting drop in demand as a response to higher prices.

Perhaps the most interesting feature of the EWG graph is a dotted line extending up to the right, which illustrates projected demand if past trends should continue. That line is based on the World Energy Outlook report for 2006, prepared by the Inter–

---

[50] M. King Hubbert, "Energy from Fossil Fuels," *Science Magazine*, February 4, 1949, vol. 109 pgs. 103-109.

[51] Energy Watch Group, "Oil Report," February 2008; www.energywatchgroup.org/Oil-report.32d637ble38d.o.html

national Energy Agency (IEA).[52] The dotted line extends ever upward, while the projected supply of petroleum falls away below. This is an impossibility, for we cannot have our cake and eat it too.

Obviously, based on these projections, that projected demand cannot and will not be filled and the gap will widen surprisingly fast, one might say like the gaping jaws of a tiger.

It should be noted that the IEA is an intergovernmental agency representing rich nations. There have been suggestions that it tends to overstate future oil supplies. In fact, according to an article in The Guardian newspaper concerning charges made by a whistleblower from within the organization, "a senior official claims the US has played an influential role in encouraging the watchdog to underplay the rate of decline from existing oil fields while overplaying the chances of finding new reserves."[53]

Keeping that caveat about the IEA in mind it is significant that in August, 2009, the British newspaper *The Independent* published an interview with Dr. Fatih Birol, the organization's chief economist,[54] concluding that: "The world is heading for a catastrophic energy crunch that could cripple a global economic recovery because most of the major oil fields in the world have passed their peak production, a leading energy economist has warned."

Although Dr. Birol hedged on admitting that Peak Oil may

---

[52] The WEO 2009 report, issued on November 10, 2009, predicted continued increases in demand. A press release stated: "Fossil fuels account for 77% of the increase in world primary energy demand in 2007-2030, with oil demand rising from 85 mb/d [million barrels per day] in 2008 to 88 mb/d in 2015 and 105 mb/d in 2030."

[53] "Key oil figures were distorted by US pressure, says whistleblower," *The Guardian.co.uk*, November 9, 2009.

[54] "Warning: Oil supplies are running out fast," by Steve Connor, Science Editor, *The Independent*, online edition, August 3, 2009

have already been reached, he indicated cause for concern and admitted that prior supply estimates had been wrong. His statements to *The Independent* are worthy of close examination. Here are further excerpts from the article:

> ... Dr Birol said that the public and many governments appeared to be oblivious to the fact that the oil on which modern civilisation depends is running out far faster than previously predicted and that global production is likely to peak in about 10 years - at least a decade earlier than most governments had estimated.
>
> But the first detailed assessment of more than 800 oil fields in the world, covering three quarters of global reserves, has found that most of the biggest fields have already peaked and that the rate of decline in oil production is now running at nearly twice the pace as calculated just two years ago. On top of this, there is a problem of chronic under-investment by oil-producing countries, a feature that is set to result in an "oil crunch" within the next five years which will jeopardise any hope of a recovery from the present global economic recession, he said.
>
> In a stark warning to Britain and the other Western powers, Dr Birol said that the market power of the very few oil-producing countries that hold substantial reserves of oil - mostly in the Middle East - would increase rapidly as the oil crisis begins to grip after 2010.
>
> "One day we will run out of oil, it is not today or tomorrow, but one day we will run out of oil and we have to leave oil before oil leaves us, and we have to prepare ourselves for that day," Dr Birol said. "The earlier we start, the better, because all of our economic and social system is based on oil, so to change from that will take a lot of time and a lot of money and we should take this issue very seriously," he said.

From all of this we can conclude that the IEA continues to

predict rising demand, but is now admitting the likelihood of falling supply. This is an economic analog to the Newtonian paradox of an irresistible force meeting an immovable object. You can't have it both ways.

A basic tenet of economics is that growing demand, unless met by rising supply, will force costs to rise. We saw that effect in action during 2008 when oil prices spiked to $147.27 a barrel. That shook the foundations of the world economy and the price run-up receded only as demand fell due to the ensuing economic collapse.

We need oil if we are to continue down the dead end path of industrial technology—but what cost is acceptable? In 2008 we learned that the world economy was not yet ready for oil prices in excess of $100 per barrel. In fact, judging from the accompanying recession, the true end of which (if any) may still lie far in the future, at those prices demand for oil cannot be sustained. Because oil is essential to industrial production, the alternative is economic stagnation or decline. Depending on your point of view, this may be a good thing.

Another point to consider about resource peaks is that even as production declines, costs increase. In the case of oil, the low hanging fruit of easily pumped reserves is gone and the remaining resources on the backside of the peak are the more remote, the more expensive to drill from the Earth, the more difficult to transport, and often of lower quality. So there's another reason to expect higher prices as resources pass peak production. Those costs cannot be escaped and will have a strong and continuing effect on entire economies and in particular on food security.

A Mountain Range of Resource Peaks

As if the emerging realization that we have passed peak oil with its attendant economic cost is not enough, consider that there are many more resource peaks looming over world economies. For example, between 2003 and 2007 copper prices rose dramatically, from around $2000 a tonne (1000 Kg.) to about $9000. It was widely believed that the price increase signaled a production peak. In 2008 prices dropped back to around $3000 in response to the economic recession, but soon resumed climbing. As of December 23, 2015, they stood at $4680 per tonne.

Copper is one of the most important mineral resources, and although it has been in use for 10,000 years, more than 95 percent of all the copper ever mined and smelted has been extracted since 1900.

Tin, lead, zinc, platinum (used in fuel cells) and lithium (essential for Lithium-ion batteries) are also believed to have peaked or are expected to peak in the near future. Peak gold was probably passed in the year 2000, since worldwide production of the precious metal has dropped by about one million ounces a year since that time.[55] In late 2009 the falling supply of gold launched a price run-up that saw several nations including India and China beginning to increase their holdings. By 2015 demand for gold had reached epic proportions. Strangely, more of the precious metal was "owned" as paper, not the actual gold, creating an illogical glut of the resource caused in defiance of the relationship of actual supply to demand.

And it's not just the major chemical elements that are at risk. According to a recent news report, the world's supply of rare earth metals is also running short. Have we reached Peak Neodymium? Peak Terbium? Peak Dysprosium? Peak Lanthanum? Strange though it may seem, these seldom mentioned elements are critical

---

[55] "Barrick Shuts Hedge Book as World Gold Runs Out," *Telegraph.co.uk*, November 11, 2009. The article quotes the president of Barrick Gold, a leading producer.

to the technologies behind hybrid cars, high-capacity batteries, efficient electric motors, and a host of other cutting-edge devices, the darlings of industrial "progress."

According to Reuters,[56] "...worldwide demand for rare earths ... is expected to exceed supply by some 40,000 tonnes annually in several years" unless new sources are developed.

China has been the leading supplier of rare earth metals, and with the growth of the Middle Kingdom's own production of high-tech products, it recently began withholding the scarce minerals from the world market, keeping them for their own use. According to the news report, Japanese companies were desperate to develop new mines in Canada's Northwest Territories.

And here is another one: Helium, and in particular in the form Helium-3. From an article in *Science* magazine we learn that He-3, an isotope of the second lightest element, is in desperately short supply. According to the article,[57] the cost of Helium-3 jumped from a previous range of $100-200 per liter to around $2000 per liter, sure evidence that another peak has been reached. He-3 is important to modern technology. Large quantities of the element have been used by the U.S. Department of Homeland Security in nuclear materials monitors. He-3 also has important applications in medical equipment and other devices, and is critical to the possible development of nuclear fusion.

The shortage has caught many researchers off base, and with the major supplier, the U.S., no longer allocating He-3 beyond its borders, many projects are in limbo. For example, after investing a

---

[56] "As hybrid cars gobble rare metals, shortage looms," *Reuters* news service, August 30, 2009.

[57] "Helium-3 Shortage Could Put Freeze On Low-Temperature Research," *Science*, 6 November 2009, vol. 326, pgs. 778-9.

billion and a half dollars in its Japan Proton Accelerator Research Project (J-PARC), Japanese scientists expected to use more than 100,000 liters of He-3 to make the accelerator operational. The article quoted J-PARC's Masatoshi Arai: "If we cannot get helium-3 … we cannot perform sufficiently good experiments." In 2015 J-PARC was working to find a suitable alternative to He-3.

Resource Peaks Bode Ill for Agriculture

These are a few examples of the many resource peaks that are jumping up in front of us almost everywhere we turn. What does all this mean for industrial agriculture? Well, everything really, for as much or more than other segments of the economy, farm technologies are based on all these and other resources. First, soaring energy costs mean rising on-farm expenses, higher cost of manufactured and processed inputs, and increased transportation expense. The unfavorable supply and cost picture for metals impacts farming as well as every other facet of the economy. And there are other resources that are of particular importance to technological agriculture, and without which it cannot continue to function in its present forms.

For example, agri-businesses use natural gas as a source of methane from which to make nitrogen fertilizer, a key requirement of growing crops. Industrial-scale farming often requires crops to be harvested wet, and farmers often use natural gas or propane to dry their grain for transport and storage. Without drying, the grain would quickly spoil unless ensiled, which would cancel its value in the open market.

Although there may still be a fair supply of natural gas, a peak may be near. In 2013 the Energy Watch Group predicted that world natural gas "will peak around or even before the year 2020." This could add to a general worldwide oil and natural gas energy crisis. In the wake of rising petroleum costs, it's natural that more energy users are switching to gas, increasing the pressure of demand on supply and price. Rising demand raises prices and hastens Peak Gas.

Phosphorus (P) is another essential ingredient in industrial agriculture, providing one of the three most important nutrients required by growing plants (the others being nitrogen and potash or potassium). Phosphate is a mineral mined from the Earth, and current estimates are that world supplies will be completely depleted by sometime in the next 60 to 130 years.[58] Peak production will occur far sooner, and may have already been reached if prices are an accurate indicator. The cost of rock phosphate rose by 800 percent in the two-year period to mid-2008, and although it subsequently declined it's estimated that Peak Phosphate will be reached in less than 25 years.

The situation may be more serious than that. A graph[59] shows actual production (black dots) plotted against a theoretical Hubbert curve that peaks in about 2035. It's worth noting that the actual figures hint a peak was actually reached some time ago and the yield is stagnant. We may already have passed Peak Phosphate.

The vanishing supply of phosphate, 90 percent of which is used for fertilizer and another 5 percent for animal feed, is a tragedy related to the divide between our industrial world and Nature. Phosphorus in the soil is used by plants, which are eaten by animals, which in turn secrete the element in their urine. In an unbroken natural cycle, P is returned to the soil through excretions. In our industrial system, animal and human wastes are treated as pollutants with no attempt to recapture the P from, well, the pee.

---

[58] "Closing the Loop on Phosphorus," fact sheet from Stockholm Environment Institute, www.ecosanres.org. The paper makes the case that phosphorus should be recycled from human waste and returned to the environment.

[59] Source: phosphorusfutures.net.

And what about that other important plant nutrient, potash (potassium)? Same tune, slightly different words. According to a news report, by April, 2008 potash prices "nearly quintupled from the beginning of 2007," going from $210 a ton to around $1000.[60] As with other commodities, the price dropped in response to falling demand and the economic recession, settling in November, 2009 at about $430 a ton, or more than twice the previous level. As the recession continued, during 2015 the price settled at around $300 per ton. Again, this demonstrates that excessive prices kill demand.

Even "Renewable" Resources Reaching Peaks

As mentioned above, the principles laid down by Hubbert can be applied to resources that might seem renewable, but which through over-use become depleted. These include topsoil, fresh water, forest products, fisheries and others. It's interesting to note that Hubbert recognized that his theory could be applied to the broader world, not just oil supplies. He wrote:

> Our principal constraints are cultural. During the last two centuries we have known nothing but exponential growth and in parallel we have evolved what amounts to an exponential-growth culture, a culture so heavily dependent upon the continuance of exponential growth for its stability that it is incapable of reckoning with problems of non-growth.[61]

How serious is the problem of resource depletion? From the short-term view on which much of economic activity is based, it

---

[60] "Potent Potables: Potash," *Wall Street Journal Digital Network*, April 23, 2008.

[61] M. King Hubbert, "Exponential Growth as a Transient Phenomenon in Human History," a paper presented at the World Wildlife Fund's Conference, *The Fragile Earth: Towards Strategies for Survival*, San Francisco, CA, 1976. It can be found at www. hubbertpeak.com/hubbert/wwf1976/

may seem of little importance. Politicians look to the next election cycle, business managers to the next quarterly statement, and commodities traders to their next commissions check. All these economic movers and shakers share an attitude reminiscent of the 17th century Cavalier Poets who lived by the Latin maxim "Carpe Diem," or "seize the day". In other words, as practiced in the 21st century, grab everything you can and forget about tomorrow.

To illustrate the position of the world as it is today, facing multiple resource peaks and other dangers, it helps to try to pull back and view things from a broad perspective. Imagine a graph that that shows a generic Hubbert Curve applied to non-renewable resources, but greatly compressed in the horizontal dimension and displayed in the timeframe of a 4000-year period beginning at the time of Christ. What you will see is a very tall and narrow mountain or spike in the middle.

Before and after the brief period in which we presently exist there are no resources, and our civilization is presently balanced on the little pointy part at the top of that spike. Think about it.

Reaching "Peak Everything"

Clearly the future is going to be different for human societies, because we are traveling the dead end path that leads to Peak Everything. Yes, everything … we are at or approaching Peak Energy, Peak Wealth, and even no doubt soon, Peak People. Peak Civilization quite likely has already been left behind.

More to the point of this book, there is evidence that we are near to or may already have passed Peak Food. In June, 2009 the BBC quoted the United Nations[62] that hunger in South Asia had "reached its highest level in 40 years because of food and fuel price rises and the global economic downturn." The BBC story

---

[62] *"S Asia Hunger at 30 Year High,"* BBC News, June 2, 2009.

reporting this news added that according to the World Bank, three-quarters of the population in South Asia, "almost 1.2 billion people … live on less than $2 a day. And more than [400 million] people in the region are now chronically hungry." Worst hit were Nepal, Bangladesh and Pakistan.

The problem has refused to go away. At the end of 2015 the UN's World Food Programme estimated there were 795 million undernourished people in the world. The WFP estimated that nearly half of childhood deaths in the world were due to poor nutrition, affecting 3.1 million children per year.

To put these figures into perspective, in 1996 the World Food Summit put on by the UN's Food and Agriculture Organization set a target of reducing the number of under nourished people in the world to about 410 million by 2015. Nineteen years later little progress has been made to achieve this goal.

To add to this troubling picture, factor in such events as years of drought and water shortages in many historically important agricultural areas. It's no wonder that the world is struggling to meet previous food production levels.

Forget Peak Oil—Peak Food could be here now, and there's more bad news to come on this front. Why do so many things seem to be going wrong, and why so quickly? Until recently when we looked around from the perspective of rising economic activity, everything looked fine to most people.

Beyond wonderful, in fact, for those who were invested in "the game," with billions to be "earned" and stuck away in numbered accounts or squandered on useless fripperies. The myth of "progress" and faith in the concept of endless growth could keep the wheels turning—but not forever, no not even close.

* * * * *

The Failure of Straight-Line Thinking

It may seem hard to understand how so many people can cling to the belief in an ever-better future. One explanation lies in the phenomenon known as linear, or straight-line, thinking. Many people tend to look at the past (and being largely ignorant of history, most of them look back a very short distance) and imagine a line extending to the present. Then they extend that line to the distant future, straight as an arrow, all the way to the end of time (or at least to an era sufficiently distant that it needn't be of concern).

Those unreasonable men and women who believe the party will go on forever would like us to think that change is linear and uni-directional. Would that it were so. Such change is easy to visualize—all one would need to do to foresee the future is to look back to history and extend the trend-line forward.

But the way the world (and practically everything else in the universe) works is not, never has, and never will be in a linear or uni-directional fashion. From the macrocosmic to the microscopic, the rule is for exponential effects that can cause change to take place with alarming rapidity.

On the mega-scale, contemplate an aging star larger than our Sun. If it contains a certain threshold amount of mass it will shine quite happily for millions of years—but in a matter of seconds a tipping point is reached and that star becomes a supernova, blowing itself to bits. Without a full understanding of stellar evolution, linear thinking could never predict such an event.

On the microscopic scale, imagine a strain of *e. coli* bacteria that lives in balance with its environment most of the time, but that suddenly bursts into runaway growth due to some change in the conditions in which it lives (for example, perhaps the potato salad having been left out overnight in a warm room). Suddenly, that normally harmless *e. coli* can become a deadly threat.

These are examples of exponential change, the kind that are found almost everywhere you look. And there is another thing to notice about exponential change, and that's the fact that it often leads to sudden "tipping points" in which conditions might change from apparent stability quite dramatically, even in a brief instant.

Could the environment of the Earth have something in common with those stars that are destined to go nova in an instant of time? Those common bacteria that can suddenly become serial killers? Hard to imagine, isn't it? And yet, we should seriously consider that question because exponential change and tipping points are more the norm than the exception in the natural world.

Our propensity to think in linear terms—and to visualize the future by reference to the recent past—is based on false observations and is actually contrary to fact. Buying into a model that predicts natural forces will operate in a linear fashion is a serious handicap when trying to envision the possibilities of change. We don't yet know what it will be like to be on the downhill side of the Hubbert curves, but it will not be an extension of the past.

Here's an analogy that may help to bring things into focus about tipping points. Imagine that you have placed a pan of cold water on a stove and turned up the heat. Failing to heed the old wives tale about watched pots, you observe the water as the stove continues to warm it. Hmm, not much going on is there? Those old wives may have had something, eh? There's the pot, its contents of water just sitting there, continuing to behave as plain old water always does. Nothing much seems to be happening. If we didn't know better, we might imagine that nothing ever will.

And yet, as we well know, as the stove continues to heat the pot the temperature of the water is rising steadily. This you know; this you understand from your direct experience. So, what happens next? You know the answer—the pot quickly advances beyond the first signs of change and comes to a rolling boil, releasing a cloud of steam and causing the water to bubble frantically and perhaps even spill over onto the stovetop. Sudden and dramatic change occurs. The water has passed a tipping point at 100° Centigrade.

This provides a good analogy for what the linear, rear-looking view can so easily lead us to believe. Just as with the slowly warming pot of water, not too much seems to be going on here on Planet Earth, and yet tipping points are a real possibility.

Based on years or reporting and many interviews with leading climate scientists, environmental journalist Fred Pearce describes the danger of tipping points this way:[63]

> Nature is fragile, environmentalists often tell us. ... The truth is far more worrying. Nature is strong and packs a serious counterpunch. Its revenge for man-made global warming will very probably unleash unstoppable planetary forces. And they will not be gradual. The history of our planet's climate shows that it does not do gradual change. Under pressure, whether from sunspots or orbital wobbles or the depredations of humans, it lurches—virtually overnight. We humans have spent 400 generations building our current civilization in an era of climatic stability — a long, generally balmy spring that has endured since the last ice age. But this tranquility looks like the exception rather than the rule in nature. And if its end is inevitable one day, we seem to be

---

[63] Fred Pearce, "With Speed and Violence: Why Scientists Fear Tipping Points in Climate Change," Beacon Press, 2007.

triggering its imminent and violent collapse. Our world may be blown away in the process.

To the careless observer, straight-line thinking seems rational—but so does the idea that the Earth is flat. The truth is that things in nature do not take place in a linear fashion; only by looking at a small segment of any timeline does that false impression emerge.

But what about now, looking ahead from the dizzy heights of Peak Everything? Consider that 4000-year graph again. It's not a straight-line world ahead, is it? In fact we see that a steep downward slope lies beyond our era, not unlike the side of Mount Everest as seen from its lofty summit—and after all, it's "peaks" that we are dealing with here.

Beyond that steep drop lies a vast and endless expanse of future history during which all of the non-renewable resources upon which our civilization relies will have dwindled and disappeared.

The Myth of Eternal Growth

Belief in the creed of eternal growth; acceptance of everything said by economists, spokespersons in government or business, or from some guy in a bar; ignorance of basic principles and passive acceptance of surface appearances—in short, shallow and linear thinking in all their various guises—tell the unwary that that ascending line on the front side of the Hubbert curve should have continued to rise, right through the top of the graph, all the way off the page and beyond, ad infinitum. Talk about a sharp disconnection between perception and reality. What's up with that?

Here's what's up: The Earth has only so much to give, and when it is gone there will be no more. As mentioned earlier, the late economist Herbert Stein once said, "If something cannot go on forever, it will stop."[64] It's worth adding that thanks to the second law of thermodynamics, nothing can go on forever, not even the universe.

Our industrial civilization, built on the ill-advised practice of unrestrained resource "development," will stop. Industrial agriculture will end its frantic rush down the dead end path… and stop. Unless we do something, soon, our civilization will stop.

There is no escaping the fact that in the not-far future we must face the challenge of adapting to these changes. The only question is: How?

We'll discuss that sticky question in a later chapter, but for now it should be noted that the transition to sustainability, if even possible, will be difficult to say the least. If runaway climate change occurs, as seems increasingly possible, the cascading problems for humanity could become insurmountable.

There's a lot of lip service paid to the subject, but the problem with most environmental movements that say they are striving to "save the Earth" is that they are doing nothing of the kind. What they want is to preserve their little part of civilization pretty much in its present form. They dream of finding easy-answer alternatives that will allow them to continue to drive high-tech cars, live in heated and air conditioned houses, work at jobs making products or providing services for a familiar economic milieu, to live in an ever-expanding world of technological wonder and comfort.

Considering the hard truth of resource depletion and looming climate change, not to mention over-population, such a future is beyond impossible.

Some even believe that this rocky transition away from the

---

[64] Herbert Stein, *Slate*, Op. cit.

industrial world can present a tremendous profit opportunity. Not only don't they expect to have to make personal sacrifices—they actually want to turn a buck on the deal, by investing in wind farms, solar energy, Russian farmland and every other scheme you can imagine. This is ironic since it's in no small part the dedication to almighty profits that got us into this mess in the first place. And consider that if it is technology that's gotten us into this trouble, how can we assume that even more technology will get us out of it?

But here's some welcome news: We don't have to worry about "saving the Earth," for it will take care of itself as it always has over several billion years. The Earth has no need of us, and on the basis of our past behavior it would no doubt be better off without us.

The less welcome message is that our focus must be on saving ourselves, the very human race itself, assuming that it is worth saving. To do so we must accept that we have gone down a dead end path, that the future cannot be like the past, that we must begin to adopt completely different ways of living, and that it will require pain, sacrifice, and change beyond our imaginings.

The message that humanity has drifted apart from Nature has always been there for those with eyes to see, but few have noticed or paid attention. The English Romantic poet William Words-worth had an inkling of it when he wrote more than two centuries ago:

> "The world is too much with us; late and soon,
> Getting and spending, we lay waste our powers:
> Little we see in Nature that is ours;
> We have given our hearts away, a sordid boon!"[65]

---

[65] William Wordsworth, from "The World Is Too Much With Us," a sonnet from *Poems in Two Volumes*, 1807. Wordsworth saw the dangers represented by the first stirrings of the Industrial Revolution.

It may seem hard to understand how so many people can remain in denial about the threats that face us. It probably comes down to psychology, and the natural tendency of individuals to protect themselves from unpleasant facts.

Some writers[66] have suggested that refusal to accept the very real possibility our technological civilization is in danger of collapse is due to a process similar to the stages of grief described by Swiss psychiatrist Elisabeth Kübler-Ross.[67] Denial is the first stage in her description of the process of grieving, followed in order by anger, bargaining, depression, and finally acceptance.

We can see that gaining a state of acceptance is not an easy transformation, and who can imagine a greater cause for grief than to mourn our very planet and the injuries we've done to Nature?

If Kübler-Ross's ideas apply to how individuals react to the present state of civilization, those in denial have several stages to go. Others are more advanced. You might profit from deciding where you fit on the scale of "grieving" over the impending changes in our world.

In the stage of denial, the most hard-core resisters often claim the danger of resource depletion is vastly over-stated, that those of us who sound the alarm are nothing but Chicken Littles, conspiracy theorists, or loonies. Their heads are firmly in the sand.

Those in the second stage of anger are looking for someone to blame. Obvious targets (and not without reason) are Big Oil, power companies, foreign countries such as Saudi Arabia or China, globalization, or those ever-popular scapegoats politicians. These might

---

[66] For example, Richard Heinberg in his book "Peak Everything," New Society Publishers, 2007.

[67] Elisabeth Kübler-Ross, "On Death and Dying," 1st pb. edition, 1970.

be people you would see waving signs on picket lines, intent on finding someone to blame for the mess.

Those who have reached the stage of bargaining might pin their hopes on an almost cult-like confidence that new technologies will allow us to find and use more resources. Bargainers are always seeking "solutions," such as escape to space or some unexpected discovery that will make everything safe again, so that "progress" can continue unhindered.

I know many people who tell me they don't want to think about this subject because it's depressing. They are in the fourth stage of the process, unable to cope with the realization of the fact that our very civilization may be teetering on the edge of disaster and the possibility of human extinction.

I rank myself in the stage of acceptance, and it's only when that point is reached that one can see clearly enough to apply the powers of reason and analysis to the challenges that face humanity. Rather than being depressed by the study of this subject, I find it of intense interest—although it's discouraging to see how many are still in the earlier stages of "grief," particularly that counter-productive state of denial.

Building Walls of Denial

There are many arguments used to build the case for denial or bargaining. For example, some point out that we will always have plenty of resources because ocean water contains vast amounts of dissolved minerals, resources that could be recaptured and used by our descendants in order to continue to travel the dead end path.

Well, they are correct that ocean water contains minerals— that's why sea salt is good for you, because it provides trace elements your body needs. Salt is the major substance extracted from the seas, along with magnesium, the only elemental mineral presently capable of being produced from sea water through electrolysis.

Other elements are found in small quantities, on the order of parts per million or less, and to extract them would require the processing of huge quantities of seawater. That, in turn, would call for vast amounts of energy, which is one of the critical resources already failing to meet demand.[68]

Yes, ocean water could provide some small amounts of a few elements, but nothing to compare with the production of rich ores through hard-rock mining, and certainly nothing sufficient to support something akin to our present technological civilization with its requirements for vast amounts of resources.

Others have suggested that even common stone contains amounts of elements such as iron, zinc, copper and lead. Thus, in theory ordinary rock could be processed to yield the necessary raw materials to keep our technological civilization going.

Again there is some truth to this. It has been reported[69] that 100 tons of ordinary igneous rock such as granite contain 8 tons of aluminum, 5 tons of iron, 1,200 pounds of titanium, 180 pounds of manganese, 70 pounds of chromium, 40 pounds of nickel, 30 pounds of vanadium, 20 pounds of copper, 10 pounds of tungsten, and 4 pounds of lead.

But again, the energy required to fractionate and refine the metals would be a serious limiting factor. These elements are not concentrated as in rich ores, but bound up in a variety of chemical compounds that would have to be processed in different ways to

---

[68] I set aside for now discussion of the will-o'-the-wisp promise of unlimited fusion power.

[69] Harrison Brown, *The Challenge of Man's Future*, The Viking Press, 1954.

yield the various mineral products. A huge amount of rock would need to be quarried and transported. Because the amounts of most elements in the stone are relatively small, an enormous amount of waste material would be created and be managed somehow.

For example, to obtain just 20 pounds of copper, 100 tons of granite would have to be mined, handled and processed, and the residue disposed of—a ratio of more than 20,000 to 1, even assuming a 100% yield. It's hard to imagine the enormous cost of such an undertaking, and the amount of energy that would be required.

We have been warned of the danger of resource depletion. Just as writers such as Malthus and Ehrlich told us of the threat of over-population, wise men have warned of resource depletion.

In 1949, Fairfield Osborn, then-president of the New York Zoological Society, published a seminal book[70] in which he compared human depredations against nature with war, comparable to the world war that had recently ended. He wrote:

> "The other war, the silent war, eventually the most deadly war, [is] one in which man has indulged for a long time, blindly and unknowingly. This other world-wide war, still continuing, is bringing more widespread distress to the human race than any that has resulted from armed conflict. It contains potentialities of ultimate disaster greater even than would follow the misuse of atomic power. This other war is man's conflict with nature."

Another early warning came in 1954 from Harrison Brown, a noted geo-chemist. He pointed out that without the resources that are even now being depleted, future humans could never build

---

[70] Fairfield Osborn, *Our Plundered Planet*, Little, Brown and Co., 1948.

another technological civilization. Here are a few poignant snippets from what he wrote:[71]

> "...a collapse of machine civilization after the disappearance of high-grade ore deposits would probably be irreversible, and the world as a whole would be covered with people living an agrarian existence."

> "Much of the knowledge existing at the time when industrial civilization reached its peak would probably be preserved, taught in schools, and passed on from generation to generation. But much of it would be valueless and, as time went on, would be lost."

> "Collapse of machine civilization would be accompanied by starvation, disease, and death on a scale difficult to comprehend."

The Club of Rome Report

The call was sounded again in 1972, this time from a program sponsored by The Club of Rome titled "Project on the Predicament of Mankind." The Club of Rome is an international think tank founded in 1968 by a group of intellectuals during a meeting in the Eternal City.[72] According to their website "they came together to discuss the dilemma of prevailing short-term thinking in international affairs and, in particular, the concerns regarding unlimited resource consumption in an increasingly interdependent world."

The report,[73] written by a research team at the Massachusetts

---

[71] Harrison Brown, Op. cit.

[72] www.theclubofrome.org. The group has its headquarters in Winterthur, Switzerland.

[73] D.H. Meadows, D.L. Meadows, J. Randers, W.W. Behrens, "The Limits to Growth," Universe Books, 1972.

Institute of Technology's Sloan School of Management, presented the results of computer modeling to predict the effects of continued exponential growth of technological civilization. The results offered a clear warning of the dead end path down which we are traveling.

The computer models took into account five major trends affecting the future of the Earth and humanity: population, pollution, industrialization, food production and resource depletion. The authors drew this basic conclusion, based on the assumption that things would continue as they were at the time of the study more than 40 years ago:

> "If the present growth trends...continue unchanged, the limits to growth on this planet will be reached sometime within the next 100 years. The most probable result will be a sudden and uncontrollable decline in both population and industrial capacity."

They pointed out that their predictions were based on a business-as-usual model, but that their dire predictions could be mitigated by aggressive action. They held out the invitation to others to join them "in understanding and preparing for a period of great transition—the transition from growth to global equilibrium." They noted that "the sooner they begin working to attain it, the greater will be their chances of success."

It's sad to see that the many graphs and tables they presented predict the world of today pretty much as it is. We have succeeded in doing little if anything to push back the approaching collapse. In fact when the team published a 20-year follow-up they titled it "Beyond the Limits" to indicate their belief that humanity had exceeded the ability of the Earth to sustain human civilization.

Ten years later the subject was revisited again after 30 years. This time the authors were "much more pessimistic" about the prospects than in 1972: The original hopes that progress could be made toward a safer future seemed to be fading. In the intro–duction to the new book they wrote:

> There are already persistent economic declines in many localities and regions. Fifty-four nations, with 12 percent of the world population, experienced declines in per capita GDP for more than a decade during the period from 1990 to 2001.[74]

They reported on a study headed by Mathis Wackernagel[75] that concluded humanity had already exceeded the carrying capacity of the Earth's resources sometime in the 1980s and had overshot by 20 percent. They added:

> Sadly the human ecological footprint is still increasing despite the progress made in technology and institutions. This is all the more serious because humanity is already in unsustainable territory. But the general awareness of this predicament is hopelessly limited. It will take a long time to obtain political support for the changes in individual values and public policy that could reverse current trends and bring the ecological footprint back below the long-term carrying capacity of the planet.

Finally, the authors commented on the question of whether the original report from 1972 was "correct," a question they have often been asked by the media. (They pointed out that their goal was never to create an inflexible "answer," but to "identify possible different futures.") They continued:

> At first the voices of most economists, along with many industrialists, politicians, and Third World advocates were raised in outrage at the idea of growth limits. But

[74] Donella Meadows, Jorgen Randers, Dennis Meadows, "Limits to Growth: The 30-Year Update," Chelsea Green Publishing Co., 2004.

[75] M. Wackernagel et al., "Tracking the Ecological Overshoot of the Human Economy," Proceedings of the Academy of Science 99, no. 14, 2002.

eventually events demonstrated that the concept of global ecological constraints is not absurd. There truly are limits to physical growth, and they have an enormous influence on the success of policies we choose to pursue our goals. And history does suggest that society has limited capacity for responding to those limits with wise, farsighted, and altruistic measures that disadvantage important players in the short term.

Others have spoken up about the threats to civilization, even the occasional politician, including of course Al Gore and also Mikhail Gorbachev. In his recent book on the environment,[76] Gorbachev relates the following "joke" (it's not really funny) that had circulated among ecologists:

> Two planets meet in space. One looks ill and complains of having contracted *homo sapiens*. The other, bursting with health, replies: 'Don't worry my friend. I had the same illness, but it went away entirely of its own accord.'

Gorbachev noted that in the late 1990s a group of biologists and (apparently rogue) economists calculated the cost of nature's gifts to us—things most economists ignore such as clear water, fresh air and fertile fields to plant. The amount totaled to 33 trillion dollars per year. Gorbachev commented: "Not that nature is waiting to be paid, but is it not so that our descendants will have to raise at least that sum in order to put right the polluted environment we are leaving them."

Indeed, it is not just a polluted planet that we are preparing to leave for our descendants, but one that has been stripped of its natural wealth—the minerals, fuels, plants and animals, fresh water, fertile soil and all the many other resources through the destruction of which we built our civilization.

---

[76] Mikhail Gorbachev, "Manifesto for the Earth," Clairview Books, 2006.

All that began when our distant ancestors set down roots and became farmers, taking the first step along the dead end path. In the resource-free world of the future, later humans will have no choice but to return to a simpler way of life, perhaps not unlike that of pre-agricultural hunter-gatherers. That is, if there will be anything left to hunt or to gather. Remember this: The Earth does not need us; we need the Earth.

The mad rush to use up all the resources of our planet had its faint beginnings 10,000 years ago with the first stirrings of agriculture. Every step toward the technological, industrialized, rushing-to-oblivion civilization of the early 21st century was initiated, aided or abetted by the advance of industrial agriculture in the past century or so. Nevertheless, whether a flint-edged sickle or a $150,000 combine harvester, a stone axe or a lumber mill, on each step of the way clever human invention has steered us down the dead end path.

Agriculture gave birth to villages, villages grew into towns, towns became cities, and finally imperial nations emerged. By becoming farmers, humans grew in numbers, leaving a natural life of peace and comfort to engage in military conquest, slavery, and widespread destruction of habitats. All the long, fateful way down the dead end path we've used up more and more of what our planet originally contained until now the cornucopia, once con–sidered the source of endless plenty, may soon be empty.

It may seem surprising that most of the growth, at least in sheer numbers of people and quantities of resources used up, has taken place in little more than a half-century since the end of World War Two. During that brief time, less than a single human lifetime, we have seen the scourge of industrial agriculture expand phenomenally. It's not really surprising at all when you consider the power of exponential increases.

To understand why this juggernaut has progressed so fast, and the urgency of our situation, consider this brain-teaser: Being informed that a pond is being covered by a growth of algae, and that each day the algae doubles the area it covers, we're asked: If the algae covers the entire pond on the 30th day, on which day will it cover just one-half? The answer, of course, is on the 29th day.

If this is a metaphor for what technology and industrial agriculture are doing to the Earth, and if we look ahead from the perspective of many resource peaks, we have reached that 29th day.

The recent period of exponential agricultural expansion is characterized as the "Green Revolution," a burst of scientific "advances" riding on a wave of fossil fuels. Ironically, the Green Revolution is generally hailed as a triumph of humanity, a feat that is claimed to have "saved" billions of lives from famine.[77] It's seldom pointed out that the supposed beneficiaries of this "miracle" are living at the very bottom of a different kind of "food pyramid," a stratified economic structure with well-fed "haves" living comfortably at the top and the poorest struggling to survive at the bottom.

Let us ponder those "saved" individuals, their lives made possible by temporary food surpluses churned out by onrushing agricultural technology based on resource destruction. They are the very poorest among us. Many are already living hungry, desperate lives. As Peak Food passes, they stare hopelessly into the gaping chasm of famine. No just and reasonable world could consider these billions as fortunate beneficiaries.

---

[77] For example, according to the website of the Norman Borlaug Heritage Foundation, "[Borlaug's] legacy includes billions of lives saved from the misery of starvation…,"www.normanborlaug.org. Borlaug was a leader of the Green Revolution who died September 12, 2009. He was, of course, a fine and honorable man who strove to perform good works throughout his life. He holds no blame; the blame lies with industrial agriculture itself. We all are its victims.

If we choose to take comfort from the appearance that we have used only half of the resources of our planet, we must ask ourselves: What will things look like when we wake up tomorrow on that 30th day?

David L. Brown

Chapter Four

Taking the Green Path

The roots of agriculture go deep, penetrating through the dark loam of pre-history to the time 10,000 years ago when horticulture, herding and farming began to replace hunting and gathering as ways for humans to make a living from nature.

For most of its history agriculture was relatively benign. Without the tools of the Industrial Revolution, farmers could not bring massive harm to the environment. Without billions of people swarming over the planet, Nature still stood a chance against us. Land was abundant, and as long as overshoot had not occurred, in many places farmers were able to let fields go fallow and rotate their crops to freshen and renew the soil. Animals remained an integral part of farming, so the cycle between the plant and animal kingdoms remained unbroken.

It was with the coming of technology and the application of fossil fuels that agriculture really began to get up to speed. That period emerged a little over 200 years ago when coal and later oil and gas were tapped to perform work formerly done by human or animal power. Renewable energy sources had been used for centuries if not thousands of years, to turn windmills or water wheels. But there are limits to what can be done with wind and water.[78] Non-renewable fossil fuels and machine technology allowed production to go beyond those natural limits.

The keystone innovation of the Industrial Revolution was the perfection of the steam engine by the Scottish engineer James Watt. He learned to turn coal—formerly used mainly as an alternative to wood for home heating, blacksmithing and other low-tech uses—into a new source of energy that could do work. His engine achieved this by heating water in a boiler then capturing the power of the expanding steam to drive pistons or turn shafts.

Drawing energy from fossil fuel allowed Great Britain, the coal-rich nation where all this madness began, to become a major economic power. Industrialization fueled the creation of a glorious empire on which it was said that the Sun would never set.[79]

The technological cat was out of the bag and it wasn't long before what some call the "second industrial revolution" was in full swing. Sailing ships became obsolete, no match for coal-fired steam ships descended from examples such as Robert Fulton's Clermont (1808). Horse-drawn carriages and canal boats were replaced by smoke-belching trains such as the Rocket of George and Robert Stephenson, which in 1829 attained 12 mph hauling a load, and an astounding 29 mph running light.

By the late 19th century coal-fired generators began to deliver

---

[78] Those who envision a future civilization powered by renewable resources may take a lesson from this fact, especially as concerns wind power. As streams and rivers dry up, there will be little future use of water power. Compared with the energy contained in a barrel of oil, wind and water are weak sisters indeed. Wind generators are not somehow "natural," but ultimate products of the machine age.

[79] Needless to say, the resources of coal were limited and despite the North Sea oil and gas discoveries that have now largely been used, the Sun has set on Great Britain's world domination. As the 18th century poet and artist William Blake predicted, "Empire is no more."

electricity. In 1882 Thomas Edison built the first practical coal-fired power station, selling electricity to wealthy residents of New York City. The new form of power was used mainly to replace gaslights, which in turn had superseded oil lamps, which had replaced candles, and so forth back to the first campfires in the wilderness.

After the turn of the 20th century steam power began to fade as the internal combustion engine emerged as the sine qua non of fast-track industrialization. The marriage between petroleum and the machine would drive civilization into our present era.

But today the oil-machine partnership is deeply troubled. The industrial world demands ever more petroleum, yet with Peak Oil upon us the Earth can no longer oblige.

In 1980 the futurist Alvin Toffler wrote of three phases of civilization, two past and one he predicted for the future. He termed these eras "waves." He recognized the emergence of agriculture as the first wave and the Industrial Revolution as the second. He proposed that the world was entering a third wave, represented by sweeping social and technological changes that he believed would result in a new and exciting "post-industrial" civilization.[80]

More than 30 years later it seems doubtful that his vision has come forth as the kind of world-changing event he anticipated. Yes, we are using more technological tools — calculators instead of slide rules; PCs instead of typewriters (and without secretaries to operate the typewriters); robots to replace humans on assembly lines; and advances in plant genetics to vastly increase farm efficiency. Toffler predicted many changes such as these, and there is no doubt that many have come to pass. But have these incre-

---

[80] Alvin Toffler, "The Third Wave," William Morrow and Company, 1980.

mental changes ushered in a new "wave," something to be compared with the world-changing invention of agriculture and the era of industrialization? I suggest not. The positive changes Toffler foresaw (and many others, some of them extremely negative, that he did not) have merely accelerated the rate at which industrialization has carried us further down the dead end path.

It should be noted that the replacement of human labor with the machine has not led us into some new era of leisure and enjoyment. Those who are unemployed or under-employed are balanced by those who work 80 or 100 hours a week just trying to keep their heads above water.

Perhaps Toffler failed to recognize that his "waves" were unnatural events, rather than laudable and beneficial stages of "progress" through which human achievement is generally heralded.

In reality his third wave is nothing more than an accelerator pedal, propelling us from a plodding step-by-step pace to a careening, breath-taking, rocket-powered rush toward the time when ... yes, the time when it will all have to stop.[81]

Futurists are fascinating people, the modern day equivalents of seers or fortune-tellers, technological mystics if you will. As far as I know they do not examine the patterns of leaves in teacups, study the entrails of animals, or peer into crystal balls. Neither do they scan the skies for astrological significance. Nonetheless, futurists are today's versions of Nostradamus.

But just exactly what is it that they do? I have some small

---

[81] Now that would be a true "third wave," a future era in a time when the industrial wave has broken on the rocks of resource depletion and agriculture will cease as an industrial war against nature. What will that be like? Toffler did not address this grim possibility.

insight into that. About 35 years ago I attended a conference at The Broadmoor near Colorado Springs. The keynote speaker during the first morning was a noted futurist whose name I have long since forgotten. That afternoon some of us took advantage of the chance to tour the nearby headquarters of the North American Aerospace Defense Command (NORAD).

During the bus ride to the underground base inside Cheyenne Mountain I found myself sitting beside the futurist. I began to ask him questions, and being of a curious nature, my process of conversation often has more in common with a probing journalistic interrogation than merely passing of the time of day.

I asked him how one goes about being a futurist; what were the methods of futurology. He hemmed and hawed for a while, but I persisted. Finally he let it slip. He said something like this: "What a futurist has to do is to look at what's happening in Scandinavia, and whatever that is, it's eventually going to come to the United States. Once you see what's happening there, you can predict with fair certainty that it will happen here."

I'm sure that was a vast over-simplification, although we can see evidence of its effectiveness in present social and economic trends in America. But still it hints to me of futurism as a blunt tool incapable of seeing the broad picture and long-term events.[82]

Once again we see an example of straight-line, short-term thinking, without consideration of possible tipping point events. As we saw in the last chapter, linear thinking isn't to be trusted, especially in the longer term. And the world can change even in

---

[82] Looking at this from the point of view of late 2015, it does not portend a bright future, for the progressive Scandinavian nations are sinking under the weight of waves of Muslim immigrants. Does this predict the future for the U.S. as well? That long-term perspective of the last century gave us a picture of ever-improving wonder, but now that one-time future is here and now, things don't look so bright.

the very short term, as it did for Saxon England on October 14, 1066, and for America on December 7, 1941 and again on September 11, 2001.

To my mind, futurists have to do better than that. Their followers do not buy their books or attend their lectures to discover tomorrow's weather or learn who might win the Super Bowl. No, we expect more. We want to know how long-term trends will affect our lives for years to come, and those of our children and grandchildren. Nostradamus understood that, which is why he is still popular after hundreds of years.[83]

In their defense, futurists are by the nature of their subject in something of a bind. The future is impossible to know except in the broadest sense, and yet their audiences expect and demand real, hard, indisputable facts, predictions that they can literally take to the bank. Perhaps more than anyone, futurists must fear the future, because it will provide ultimate judgment of their life's work.

Now in the interest of full disclosure, in case it has not already become evident I am myself engaging in the examination of possible futures. I am engaging in research, weighing facts, describing possibilities, and reporting my conclusions in the clearest and most engaging way I can. And, yes, I am filtering everything through my opinions, built up over decades of experience. No one is immune to that.

To further state my case, there are several important differences that set my approach in this book apart from that of a

---

[83] Nostradamus was also clever enough to couch his predictions in vague, poetic forms that could allow numerous interpretations. Many futurists could take a lesson from that. Remember what can occur when you make predictions that are too specific, as Paul Ehrlich did in *The Population Bomb*. It's better to provide evidence and raise questions than to make rock-solid assertions. You can be 100% right in a broad general sense, but see your argument fail because you carried it too far.

practicing futurist. First, I am aware of the error of linear thinking and always take it into account, viewing the world through the lenses of rising and falling trends, tipping points, and inevitable limits. There are infinite variations in Nature and in the events of history. Because there is no such thing as certainty in thinking about the future, we can only suggest possible ways in which the trends of the past and present may play out, based on the facts as they are.

Second, I try to take the longer view, for example by focusing not on what happened in Sweden last year or Norway last week, but on what may have occurred in the Fertile Crescent 10,000 years ago; archaeological discoveries in Mesoamerica; events in Renaissance England; politics in the mid-20th century. I take very seriously George Santayana's warning: "Those who cannot remember the past are condemned to repeat it."[84] It is my conviction that a lack of connection with history is one of the worst of human failings.

Third, I take a broad view, striving to draw facts of the real world from many sources, sorting out the most rational of them, and only then drawing possible conclusions. I avoid buying into sterile, simplistic, one-track scenarios that emanate from the ivory towers of the world, those isolated halls of the Academy and think tanks inhabited by economists, historians, sociologists, psychologists, political scientists, and philosophers, each firmly entrenched in his or her professional rut of "soft science," largely unaware of the realities beyond their specialized fields.

And yet, I do not reject any source, but apply Ockham's

---

[84] George Santayana, "The Life of Reason," Vol. 1, 1905-6. This statement is often misquoted in various versions.

Razor,[85] a dollop of reason, and as much plain old horse sense as I can muster to search out cross-connections between a multitude of facts and ideas. Every thinker, no matter how wrong-headed, has something to add, even if only to provide negative evidence through patent absurdity, like the man who believed the Earth could support so many people that I would have more than a million in my house.

Finally, although my opinions are mine alone they are based on a lifetime of reading, seeing and probing. As an agricultural editor and journalist I have interviewed several thousand farmers and agri-business people during my life, and personally visited farms all across North America and in Europe, Australia, New Zealand, Thailand and China. I've written more than a thousand articles about various phases of agriculture, and even a complete farm management handbook.

How will my speculations turn out? Don't know. Let's get together over coffee in fifty or a hundred years and compare notes. Meanwhile, my present intention is to present facts and informed opinion, stimulate thought, entertain, and illuminate what may be the most critical subject in human history. I invite you to ponder the points I raise and devise your own picture of things to come. Yes, you too can be a futurist!

In that spirit, it is my personal conclusion that Toffler's "third wave" has come, and gone, and that it turned out to be pretty much more of the same. It has yielded no historic turning point, no post-industrial golden age, but merely the same old unsustainable technological civilization on steroids and rocket-powered roller skates. It's the downhill, back side of the Hubbert resource curve.

---

[85] The English philosopher William of Ockham (1285?-1349?) proposed his "razor" which holds that when competing explanations are present, the simplest is most likely to be true. This concept is faintly reflected in the modern-day expression known as KISS, for "keep it simple, stupid."

Certainly, if Toffler thought it would be, the future that's emerged since his book appeared cannot be viewed as an exciting new era. We see no evidence that we are living in a golden age of civilization, nor that one may lie just around the corner. In fact, as I suggested in the last chapter, we have no doubt passed Peak Civilization—perhaps around 1776 when James Watt built the first of those steam engines that changed the world. (Umm, let me think—did anything else happen that year?)

The Post-War Burst of 'Development'

Nowhere has this acceleration of "development" been more evident than in agriculture, and especially during the period of transformation that began soon after the end of World War Two and culminated in the intensive industrial farming of today. This has been a period of soaring food production through unnatural means, yielding a temporary and unsustainable glut of food that has carried human numbers to new heights. As we have seen, that rapid expansion has been achieved through the depletion of resources, especially oil. In less than the span of a single lifetime it has propelled us ever further down the dead end path.

Let's take a closer look at the recent period of unprecedented agricultural transformation that took place since 1945. When discussed in the context of the "developing" world, this period is generally called the Green Revolution. The term was coined in 1968 by William Gaud, then director of the United States Agency for International Development (USAID), who said:[86]

> These and other developments in the field of agriculture
> contain the makings of a new revolution. It is not a
> violent Red Revolution like that of the Soviets, nor is it a

---

[86] Speech given before a meeting of the Society for International Development, March 8, 1968, Shoreham Hotel, Washington, D.C. Source: Wikipedia.

> White Revolution like that of the Shah of Iran. I call it
> the Green Revolution.

It is interesting to note the examples he used for comparison. The Shah of Iran fell from power in 1979 and the wheels dropped off the Soviet Union in 1991. What date will history mark as the end of the Green Revolution? I will leave that question to the futurists, or to you, my reader.

We must recognize that at the time this "Third World" initiative was taking place, the same forces were at work in the rich world, and especially in the United States. Very much at work.

No, the Green Revolution didn't happen just in the poor nations, to "help" the less advantaged. It happened everywhere, and especially it happened here where industrial farming was applied with a vengeance. America had emerged from the World War as a major power, both militarily and economically. She had it all, and especially oil, for remember that in the 1940s we still had nearly a quarter century to go before Hubbert's Peak would arrive in America.

The country was awash in the stuff—cheap, plentiful, and ideally suited to fuel the greatest period of technological expansion in history. It carried us in a blink of time from when my grand-father was farming with a team of horses to Neil Armstrong's "one small step" on the Moon. Industrial agriculture was advancing like a tidal wave, transforming food production around the world.

One of the primary methods of the Green Revolution was to "improve"[87] plant genetics, and that feature has gotten most of the

---

[87] I put quote marks around this word to indicate it is a brother or sister of "progress," "development," and those other code words that are used to make the systematic destruction of our environment sound like a really good idea. These "improvements" were largely a matter of adapting crop varieties to the

good press. But more importantly, as with industrialization in general the Green Revolution was based on a platform of cheap and abundant oil. Just as medieval alchemists strove to transform lead into gold, we learned to turn petroleum into food—by using it to produce fertilizer and farm chemicals; to build a vast worldwide transportation network; to power rumbling fleets of machines to till, plant and harvest crops from the soil; and to create a global food processing and marketing system that could have been devised in the sinister laboratory of Doctor Frankenstein.[88]

Should it be a surprise to us that what we call the Green Revolution—I'm talking now about the version that began with improvements in wheat production in Mexico and spread to programs in India, Africa and elsewhere around the globe—was largely funded by the Rockefeller and Ford Foundations? It shouldn't be, because as I recall the Rockefellers had something to do with oil, and wasn't Ford a major producer of farm machinery? Well, yes.

Remember that the latest manic, grab-everything-and-forget-tomorrow phase of industrialization came about as the result of the marriage between oil and the machine, a mating that gave fruit to self-interested economic entities such as those of the Rockefellers and the Fords and many lesser examples.

Are we to imagine that these individuals and organizations stepped forward to create the Green Revolution out of their undiluted generosity toward their fellow beings—or did they have

---

methods of industrial agriculture, making them ready to thrive in an artificial environment.

[88] Fun quotes from the 1931 movie *Frankenstein*: "Crazy am I? We'll see whether I'm crazy or not." "Now I know what it feels like to be God!" "You have created a monster and it will destroy you." I mention these because they could perhaps apply as well to the development of industrial agriculture as to the animation of a monster.

something else in mind? Something like creating global markets for their products?

Perhaps the most basic method of the stage magician is the application of misdirection. The performer makes a busy fuss over *here*, so that you don't notice him depositing a rabbit into the empty hat over *there*. Thus, when he produces the rabbit the audience is suitably amazed and delighted.[89]

Charitable foundations come in all shapes and forms; many are actually dedicated to doing "good works" as they see them. But we must always be on guard lest misdirection could be involved. Yes, Virginia, seemingly humanitarian organizations may exist that do things over here to distract our attention, while over there they stuff entire generations of rabbits into a multitude of hats for their own benefit.

The Emergence of Profit as the Ultimate Good

We must never forget that for many people engaged in business, and virtually all corporations, profit is the most important thing that could ever be, more cherished than pride, more valued than wisdom, more precious than the future of the entire planet. Something to keep in mind isn't it?

And thus did the Green Revolution have its beginnings, wrapped in the clothing of humanitarian good will but driven by the desire of those "unreasonable men" of G.B. Shaw—opportunistic individuals and the corporations and foundations they control—to change the world for profit while in the process using up the

---

[89] The members of the audience are known in the trade as "suckers," and many years ago showman P.T. Barnum is said to have estimated that one was born each minute. At the present rate of population growth that process is undoubtedly taking place much faster. There are few statistics to verify it, but today's pool of potential suckers is no doubt immense.

irreplaceable resources of the Earth with little if any concern for the future.

"**PROFIT NOW!**"—it could be a bumper sticker for executive limousines. Grab the money and let the poor, the honest, the hard working, yes let me say it, the "suckers" of the world and their descendants for all time worry about the future—that seemed to be the general drift of 20ᵗʰ century food and energy entrepreneurship. It has led us to the present time when the entire portion of the world economy devoted to food is in the hands of a few giant corporations, purveyors of seed, fertilizer, fuel, machinery, credit, heavily processed foods—all manner of things required for the unnatural monstrosity that agriculture has become.[90]

This transformation was woven from many threads. To put things into perspective, we need to briefly note that in the last century agriculture was starting to hit a wall. We'll discuss the critical subject of soil in a later chapter, but you should be reminded that much of the Fertile Crescent in present-day Iraq—the very birthplace of agriculture and perhaps the model for the Garden of Eden—is more or less a howling desert today.

By the 1920s and 1930s most of the virgin land of the planet had been put to the plow, and much of that had already been seriously degraded or even destroyed. Land that had once brought

---

[90] Let me make it clear that I am sympathetic to capitalism, but only if conducted in a moral framework. Profits are not in themselves evil, but evil people and enterprises often seek profits through improper means. Moral capitalism must consider the effects of all actions taken in the quest for profits, and should avoid causing harm to anyone, whether directly or indirectly. Capitalism that thrives on the destruction of the environment and the misfortune of the many victims of its slash-and-burn tactics cannot be allowed to exist in a sustainable future. The principles behind "moral capitalism" were first set down by Aristotle and embraced by the 20ᵗʰ century philosopher and novelist Ayn Rand. In short, capitalism should abide by the Golden Rule. If it does not, it constitutes crime against humanity.

forth rich harvests no longer did. Everywhere the soil was sick, for
its nutrients had been sucked out by too much continuous
cropping, too much of the plow, too little of dedication and
concern for the future. The value of the land was being squandered
through over-use and erosion. The planet was filling up with
people, so as the soil became worn out and barren, farmers no
longer had the option of migrating from England to Massachusetts
…to Ohio…to Missouri…to Oklahoma…to California, leaving
behind land that had been logged over or burned, ripped up by the
plow, abused, depleted of its riches, and wasted as if there were no
end to its bounty.

A Close Encounter with Overshoot

Having spread across the New World, humanity was at the
brink of overshooting the Earth's carrying capacity. By the early
1930s most land could not yield very much, not nearly on the scale
needed for the population growth that was taking place. For
example, corn production in Indiana was around 20 bushels per
acre.

In 2014, under industrial methods, it came in at an average of
188 bushels[91] and many farmers would be ashamed to harvest less
than 200 bushels.

The consequences of allowing the land to become sick were
seen around the world. The Dust Bowl of the 1930s was one result.
In other regions, great famines occurred in the most over-
populated areas of the world. In India an estimated 3 million
people died during the famine of 1943. The Great Chinese Famine
from 1958 to 1961 is estimated to have caused the deaths of as
many as 38 million. The Soviet Union suffered chronic drought

---

[91] "Indiana Sets Record for 2014 Corn and Soybean Yields," *Hoosier Ag Today*
website, January 12, 2015.

and famine until a panic in 1963 caused the country to give up on trying to be self-sufficient in food and begin importing grain. Altogether there were more than twenty identified famines during the first half of the 20th century.

With soils used up, eroded, washed or blown away, thinned and impoverished through over-use, the human population was bumping hard against the ability of the Earth to support it. Was it a coincidence that during the 1920s in America it was farmers who first fell into economic distress, and that the Great Depression and worldwide conflict were soon to follow? The cornucopia was beginning to lose its power to provide.

What to do? It seemed that humanity was reaching Malthusian limits. We had overshot the Earth's capacity to sustain us. The future looked dim—and yet already a great turnaround was beginning to take place.

The problem was that the soil was poor. The solution: Fertilize it. And not with ordinary animal manure and crop rotation with legumes as farmers had done for millennia—this new program was based on the use of unnatural artificial fertilizer, produced with the aid of fossil fuels.

As we saw in the last chapter, three elements—nitrogen (N), phosphorus (P), and potassium or potash (K)—are the major nutrients required by growing plants. The king of them all is nitrogen, and it's also the most elusive. Phosphorus and potash can be mined from the Earth, at least for now, but even though the air we breathe is 80 percent nitrogen, atmospheric nitrogen is inert and cannot be used by plants.

Nature has a way of dealing with this, and it worked for hundreds of millions of years before human beings came along. She did it through natural cycles from which nitrogen was "fixed" (that is, turned into chemical compounds that plants could use) by bacteria in the soil and in the guts of animals. In addition, a certain family of plants known as legumes provided themselves with the vital nutrient by becoming hosts to nitrogen-fixing bacteria. For hundreds of years farmers grew legumes as "green manure" crops, plowing them down to enrich the soil.

Now those legumes, mainly soybeans, have been transformed into unhealthy and unnatural—but highly profitable—food for humans, not for the soil from which they come.

Better Living Through Chemistry — Not!

It all began to change in 1913 when a German chemist named Fritz Haber found a way to fix nitrogen from the air by combining it with hydrogen to form ammonia, or NH_3. The Haber process requires a lot of energy. It also uses quantities of free hydrogen, which is generally provided by the methane in natural gas.

Haber's discovery was important because … well not what you might think. Ammonium nitrate is a key ingredient in explosives, and it was in no small part through Haber's efforts that the 20th century became one of the most dangerous and deadly periods in history, with two world wars and hundreds of lesser conflicts. "Better living through chemistry?" I don't think so.[92]

In the United States during World War One a dam was built at Muscle Shoals on the Tennessee River. Its purpose: To generate power to be used to fix nitrogen into ammonia. The main idea was to provide munitions for the war, but lip service was given to the

---

[92] Haber also was instrumental in developing poison gas used against the Western allies in World War One. When he told his wife, also a chemist, what he had done she committed suicide. He also played a key role in the development of Zyklon-B, the "insecticide" used to murder millions in Nazi death camps. Ironically, he was born a Jew but had converted to Christianity. He won the Nobel Prize for Chemistry in 1918 for the Haber process of nitrogen fixation.

goal of making artificial fertilizer for the impoverished South. Historian Paul C. Conkin tells the story:

> The fertilizer proposals bogged down in the 1920s with the battle over public versus private operation of the Muscle Shoals project, but by then a few commercial fertilizer companies were using the Haber process.
>
> In 1933 with the creation of the Tennessee Valley Authority (TVA), the advocates of public ownership won out. Muscle Shoals became the site of fertilizer research, development and demonstration in the National Fertilizer Development Center. In the Depression years, the TVA concentrated on developing new and better phosphate fertilizers, or what is called triple superphosphates. During World War II it produced ammonium nitrate for munitions and started selling it to farmers in 1943.[93]

Now at last industrial agriculture had the tool it needed to resume the catastrophic growth that had begun to slow as a result of soil depletion, what we might call the back side of the Peak Soil curve. By spreading fields with granular fertilizer containing compounds of ammonium nitrate, or by injecting the soil with anhydrous ammonia, a dangerous liquid form of NH_3, even the most severely depleted soils could be brought to peak production with the help of a little P and K.

But there was a problem, because the all-out application of industrialized farm production finally broke the cycle of nature completely. Whether they realized it or not, humans had now set out in earnest to utterly replace the balances between a complex web of perennial plants and animal life, ranging from buffalo and deer down to thousands of bacteria species, nematodes, earth–

---

[93] Paul K. Conkin, "A Revolution Down on the Farm: The Transformation of American Agriculture Since 1929," The University Press of Kentucky, 2008.

worms and many other interacting residents in the complex biological environment of the soil.

Where once herds grazed on fields of eternal grass, all that is gone. Gone are nearly all of the animals themselves, replaced by domesticated species crowded into "factory farms." Gone are the perennial prairies, destroyed by the plow and replaced by annual monocrops of corn, wheat, soybeans and a few other genetically modified plants with the purpose of feeding a swelling overburden of humanity—at maximum profit.[94]

"Factory farming" has taken cattle from the land and concentrated them in feedlots. Hogs are confined in finishing barns, fed unnatural diets of corn and made to wallow in their own waste.[95] Hens roost in tiny cages and never see the Sun. Veal calves are kept in narrow crates, equally separated from Nature. Dairy cows are pumped full of antibiotics and growth hormones and stuffed with corn and soybean meal, not the grass that is their natural food.

The natural balance is broken because manure, once the precious contribution of animal life to the cycle of nature, has today become a major pollutant.[96]

---

[94] It should be noted that any profit that accrues to the farmers themselves is purely incidental, crumbs from the table of the industrial war against nature. The real money is reserved for those who deal in oil, chemicals, machinery, seed, and finance and the mega-corporations that process and market what passes for "food" these days. For many farmers, the juggernaut of industrial agriculture has meant an endless cycle of debt, insecurity, hard work, anguish, failure, and suicide. Just as animals have been separated from their connection to the natural cycle of the land, so have the farmers themselves.

[95] Contrary to popular impressions, in their natural environment pigs are essentially clean animals.

[96] "Uncontrolled Factory Farm Manure Causes Pollution and Threatens Health, According to Comprehensive New Report," News Release from Consumer's

What remains of the once-fertile soil becomes sterile and barren as its remaining natural elements are "mined" by con-tinuous cropping and the ever-increasing application of fertilizer, insecticide, herbicide, fungicide, irrigation water and all manner of other insults. Industrial methods suck the life out of the Earth's soil, leaving nothing more than a mineral amalgam on which food crops are artificially grown. The Earth's soil is treated almost like an enormous Petri dish, a mere substrate for an artificial process.

It's of particular concern that the land is drenched with insecticides. Today's farming uses whatever it takes to kill anything that threatens the sacred monocrop, whether it's corn earworms, soybean aphids, spider mites, slugs, or cotton boll weevils—chemical warfare is waged everywhere that crops are grown. And the warfare is total, because besides those targeted "pests" the poisons kill almost everything else they touch—monarch butterflies, bees that are needed to pollinate flowering plants, earthworms and soil nematodes, fingerlings in nearby streams and ponds, birds, frogs, mice and baby rabbits.

An Invisible Holocaust of Death

Look across any field where industrial agriculture is practiced. Looks peaceful, doesn't it? But, what you are witnessing is an invisible holocaust of death, the full horror of the all-out war against Nature.

The holocaust doesn't end with a scorched earth policy toward all members of the animal world. Non-crop plants are treated no better, designated as "weeds" and attacked with utter ruthlessness. Herbicides, weed-killers—whatever you call the stuff, it's more chemical warfare against Nature. The infamous Agent Orange used in Vietnam is one such evil substance.

Union, December 3, 1998. The report was jointly prepared by the Natural Resources Defense Council and the Clean Water Network.

So-called "no-till" or minimum tillage farming is alleged to be environmentally friendly because it does not involve annual plowing. What it does is perhaps worse. First the land is sprayed with a broad-spectrum "weed killer," a chemical that destroys anything that uses photosynthesis, that is, any member of the vegetable kingdom.

Once the field is barren, the sacred crop is sown among the dead remains. This is heralded as an advance because it conserves moisture and reduces costs by eliminating plowing, all while preventing any interference by Nature in the process of growing the domesticated plants favored by industrial farming. Farming has become an unnatural art.

And now through genetic engineering crop varieties have been developed that can selectively withstand the plant killers, notably those called by the brand name "Roundup Ready," because they are created through genetic engineering to be immune to glyphosate, the active ingredient in the widely used weed killer.

First to be bred for resistance were soybeans, followed by corn (maize), canola, sugar beets, cotton and alfalfa. Resistant wheat is currently under development. By 2015, 94 percent of the soybeans grown in the U.S., 92 percent of corn, and 94 percent of cotton were glyphosate resistant varieties.[97] Using these resistant crops, farmers can attack and destroy any competing plant that dares to raise its head by spraying the field repeatedly.[98]

Roundup was invented by and is a trademark of chemical

---

[97] "Recent trends in GE adoption," U.S. Dept. of Agriculture, Economic Research Service, website page updated July 9, 2015.

[98] Ironically, in 2010 it was reported that weeds were developing immunity to Roundup, just as bacteria become able to resist antibiotics. This will undoubtedly lead to the search for even more virulent plant-destroying chemicals.

giant Monsanto, which has also been the leader in development of resistant seed varieties—a perfect example of how to lock in a market. The patent on glyphosate expired in 2000, allowing other companies to produce the chemical, although Monsanto has kept the market lead.

The use of herbicide, and particular glyphosate, has exploded in the last few years. Here's a quote from the Environmental Working Group, a non-profit environmental advocacy group:

> American growers sprayed 280 million pounds of glyphosate on their crops in 2012, according to U.S. Geological Survey data. That amounts to nearly a pound of glyphosate for every person in the country.
>
> The use of glyphosate on farmland has skyrocketed since the mid-1990s, when biotech companies introduced genetically engineered crop varieties (often called GMOs) that can withstand being blasted with glyphosate. Since then, agricultural use of the herbicide has increased 16-fold.[99]

Not even the lowly fungus has a chance on an industrial farm, with a "third wave" of chemical warfare featuring anti-fungal sprays. The chemical holocaust continues unabated against all forms of life. Can that be good?

Through the process of industrial agriculture, artificially fertilized, irrigated and chemically protected crops can be grown on barren sand, as can be seen in the fields of California's Imperial Valley, essentially a desert.

"Soil? We don't need no stinking soil!" That might be the war cry of industrial agriculture. Technology appears to have tri–

---

[99] "Glyphosate Is Spreading Like a Cancer Across the U.S.," by Mary Ellen Kustin, senior policy analyst, Environmental Working Group, April 7, 2015, www.ewg.org.

umphed over nature, a clear win for humankind. Is it time for us to raise the flag of victory? Well, perhaps not.

Never mind that those crops are lacking the many micro-nutrients that come from rich, living soil, and that because of that people are sick and dying from a host of diseases that were never known when humans lived in balance with nature and ate natural foods.

Never mind that farm chemicals are polluting the environment, destroying the fragile balance that has taken nature hundreds of millions of years to create.

Never mind that industrial agriculture is a house of cards built on the delusion that natural resources will last forever, and that the technological monstrosity is about to fall down around us.

And most ominous of all for we humans, never mind that Nature always wins—or at least she always has and if she ever should lose, Planet Earth would become a barren, sterile ball of rock.

Some fear that may happen, but I personally am betting on Nature to survive. Her precious environment has already suffered grave wounds, damage that will take millions of years to repair, but I believe that she will overcome the depredations of *Homo agriculturis*, agricultural man, just as she has survived past events such as the end of the age of dinosaurs.

But whether or not there is a future for life on Earth, we must ponder whether any of our descendants will be part of that future. That question is of no concern to Nature.

I can't help but think that this "experiment" we call industrial agriculture might be the greatest demonstration ever of the law of unintended consequences. Can we really think there will be no payback for the onslaught of insults and injuries we are visiting upon the Earth? Hardly a day goes by without new evidence of just such unexpected events.

Unintended consequences generally occur out of ignorance or lack of full understanding of the complexities involved. Nothing is more complex than the web of Nature, the environment from which all life comes. On the grand scale of the entire planet, can we afford to experience that "Doh!" moment when it turns out that "Uh oh, maybe we shouldn't have done that!"

Many seem to believe that the Green Revolution has "solved" the problems of hunger and over-population. Nothing could be farther from the truth. It has only led us farther down the dead end path.

The Third World "Miracle"

Now that we've taken a peek behind the curtain at the real Green Revolution, the one that has transformed the First World, let's briefly examine the Third World version that gets all the press, the poster child for industrial agriculture. Every movement of this kind must have its heroes, and shadow operatives such as the Rockefeller and Ford Foundations certainly do not want to stand up and present themselves as public targets. Enter the late Norman Borlaug, a plant geneticist who became the very symbol of the Green Revolution.

As I touched on in the last chapter, the late Dr. Borlaug truly was a hero, a fine and dedicated man. My argument is not with him and I would never say anything against him. My argument is with the many unintended consequences of his work, and with the way it was used as a conduit to introduce industrial agriculture into the Third World.

In the late 1940s, working in Mexico for the Rockefeller

Foundation, Borlaug began developing new strains of wheat that were resistant to fungus and wheat rust and that could therefore produce higher yields.

It was a challenge. At first it worked so well that the wheat heads grew fat and caused the plants to fall over, what farmers call "lodging". Borlaug then crossed his Mexican plants with short-stemmed dwarf wheat varieties from Japan to make the stalks stiffer.

Next he worked on improving the efficiency in how the plants used nutrients, including water. Borlaug's efforts yielded a net increase in wheat yields of 2.1 percent per year between 1950 and 1990—tripling harvests where the Green Revolution plants were grown. Borlaug later focused on Asian rice with similar results, and even attempted to "improve" crops for conditions in Africa.[100]

Food production soared. Paul Ehrlich's predictions of an exploding population bomb were discredited. Even though there were billions of new human beings, they were consuming more calories than their forebears. The problem of feeding the world had been solved, hurrah! Right?

Well, no, because as we have seen the expansion of food production has been achieved only through depletion of a wide variety of irreplaceable natural resources, including but not limited to petroleum, natural gas, phosphate and potash, fertile soil, and fresh water. Those resources are peaking or have peaked, and the momentum that was begun in Mexico in the 1940s is running out of steam.

We are arriving at Peak Food, but not yet Peak People. With more hungry guests arriving at the table each day, and oil and

---

[100] A good summary of Dr. Borlaug's work can be found at the web site of The Norman Borlaug Institute for International Agriculture, http://borlaug.tamu.edu/

other resources entering decline, can widespread famine be far behind?

It's a serious question, one that has been with us for some time, echoing Malthus and Ehrlich. Besides those two there have been many voices of warning, largely ignored if not attacked by those who cannot face the fact that the free ride on a wave of petroleum may be coming to an end. Writing in 2000, environmental journalist Richard Manning reported:[101]

> From the beginning the Green Revolution has had its critics, especially those who have suggested that its heavy reliance on high inputs of water, capital, and chemical fertilizers and pesticides are simply not sustainable

> The sense of discomfort with the Green Revolution is no longer limited to its critics. There is consensus that the techniques that have brought us this far will not be able to sustain us in the future. Production is leveling off. Since 1989-90, world grain harvests have risen on average only [one-half of one] percent a year, a quarter of the rate of the Green Revolution boom years. Changed political circumstances, particularly the collapse of the Soviet Union and the resulting economic chaos in one of the world's most important grain-producing regions, offer partial explanations, but there are signs that, politics aside, Green Revolution techniques are approaching the limits of what they can produce.

> If that's true, not only will supply be constricted but the demand side of the equation will also be thrown into flux. From the beginning, agriculture has been the primary engine of human population growth; the dense package of storable carbohydrates that grains provide

---

[101] Richard Manning, "Food's Frontier: The Next Green Revolution," University of California Press, 2000.

allows mobility, cities, hierarchy, technology, medicine, longevity. We count on more agriculture to provide food for ever growing numbers of people, the solution to the population problem. We forget that the relationship is circular, dynamic, and not at all simple.

Nearly a decade on, those concerns have become of critical importance. World stocks of food have dwindled to only a few weeks supply and are likely to fall further. Climate change is causing drought to spread and melting glaciers to cause the oceans to rise and rivers and streams to dry up. And all the while those resource peaks are looming ahead of us on the dead end path.

The deniers, let us kindly call them optimists, are convinced that technology can save us. More of the same … will save us. We can work our way out of the problem by just continuing to do what got us into trouble in the first place, hands in pockets, whistling through the graveyard, confident in the way that only ignorance can yield.

Renowned editor, journalist and diplomat Clare Boothe Luce once said" "The difference between an optimist and a pessimist is that the pessimist is usually better informed.

How realistic are these optimists, trusting in technology as an ultimate panacea? Not very, I'm afraid. We hear that all we need is a "new" Green Revolution—breed plants that need less water, less fertilizer, that can resist salt and other toxins in the soil that have resulted from irrigation; create "super" plants from the planet Krypton that will allow us to leap over tall buildings and solve the threat of starvation fast as a speeding bullet. Yes, and fairies will come each night and lay out a feast for us all.

Oh, wait, there are no fairies are there? And no planet Krypton. We'll have to make do with what is real, not what we hope, imagine, or fantasize will happen. Oh, bummer.

Instead of accepting that more technology will save us from the indignity of hunger and economic collapse, we need to ask questions such as these:

> How can "super" plants use nutrients more efficiently when the cost of unnatural artificial fertilizer grows beyond the reach of most farmers, and particularly those most vulnerable in the developing world?

> How could "super" plants thrive on less water when there is not enough fresh water left? Even now in California, farmers are being cut off from water supplies so that Hollywood characters can keep their swimming pools filled.

> Most especially, how can we sustain crops of "super" plants at costs that can be afforded by seven or eight billion human beings when the once-abundant resources needed to do the job have been squandered in a burst of industrial production that is doomed to end, to stop?

Never forget that oil and food are joined at the hip. So are many other non-replaceable resources, and none of them will be available forever—or in many cases, for much longer at all.

The answers to questions such as these are critical to the future of civilization. But one thing is certain, and that is that we cannot count on being able to continue much longer down the dead end path. We cannot because the Universe is ruled by the Second Law of Thermodynamics,[102] the one that says that nothing remains the

---

[102] The second law, in various versions, explains the phenomenon of irreversibility. It describes the way in which the Universe is inexorably running down. The decline in energy is measured by the term "entropy." A version of the second law popular with engineers is "There ain't no such thing as a free lunch," often abbreviated as TANSTAAFL. The statement was popularized by science fiction writer Robert A. Heinlein, and was often quoted by the economist Milton

same; that energy can never be created but only diminishes over time; that the Universe is running down like a mechanical clock.

And yet as we have seen, many people sleep soundly each night, secure in the idea that technology will continue to improve the lot of humanity. They believe that the Green Revolution has "solved" the problems of hunger and over-population, when in fact all it has done is delay the time of reckoning.

The environmental sociologist William R. Catton, Jr. in a little-heeded book titled "Overshoot,"[103] called this a "delusion," and compared its true believers with the cargo cults of South Pacific islanders who observed airplanes delivering food and supplies to soldiers and airmen during World War Two. The cultists believed the planes were sent from the gods, and that by building their own "air fields" they could also attract the riches from the skies. They cleared "runways," built "control towers" of sticks, and sat back to await for the gifts from above. Needless to say, they were disappointed.

Drawing the comparison between cargo cultists and present-day "optimists," believers that technology will always provide answers, Catton wrote:

> The cults that won Cargoist adherents among the citizens of advanced nations were not always obviously religious. The Type II belief held that great technological breakthroughs would inevitably occur in the near future, and would enable man to continue indefinitely expanding the world's human carrying capacity. This was mere faith in a faith, like stock market speculation. It had no firmer basis than naive statistical extrapolation—

Friedman. In the future, as entropy works its way through the Earth's remaining supply of resources, there may be no lunches at all, at any price.

[103] William R. Catton, Jr., "Overshoot: The Ecological Basis of Revolutionary Change," University of Illinois Press, p.b. edition 1982.

the uncritical supposition that past technological advances could be taken as representative samples of an inherently unending series of comparable achievements.

Again, this is an example of linear thought, basing future expectations on the assumption that past trends will continue forever.[104] History has proven this false on countless occasions, but who's paying attention to history? Catton added:

> They could not even see that the high-yield grains either would hasten the exhaustion of the soils on which they were grown or would intensify agriculture's precarious dependence upon a chemical fertilizer industry.

Two and a half millennia ago the Greek philosopher Heraclitus realized that nothing was immune to change, stating: "You can not step twice into the same river."

Thanks to the Second Law, change is the only thing we can truly count on, the only eternal reality of the Universe. No matter how much we may want to hold onto the present state, Time's Arrow carries us on into a changing future.

We have no alternatives but to learn to deal with change—or to cease to be.

---

[104] There is an old anecdote that illustrates the fallibility of this way of thinking. A man falls off of the Empire State Building. As he passes the fiftieth floor someone leans out of a window and asks him how things are going. "Fine so far," comes the reply.

David L. Brown

Chapter Five

The First Revolution

Agriculture did not begin with a bang. None of its various forms—horticulture, farming, herding, animal husbandry, gardening—started with a "Eureka" moment such as the discovery of the bow and arrow or the wheel. Agriculture developed through myriad evolutionary processes over hundreds or thousands of years, finally combining to transform human societies beginning about 10,000 years ago in the Neolithic, the "new stone age."

Many creation myths touch on the transition of humans into farmers and herders, not least the Old Testament Book of Genesis. Here we learn of the "Garden of Eden," which we may view as a metaphor for the pure original environment of the Earth when human beings, as represented by Adam and Eve, lived in a natural state. Gaining the knowledge to defy Nature, the mythic couple was expelled from their former comfortable and easy life as hunter-gatherers, where they had lived in concert with all the plants and animals of the Earth.

As the creation story is told: "Therefore the Lord God sent

him [Adam[105]] forth from the Garden of Eden to till the ground from whence he was taken."[106]

We are told that Eve gave birth to two sons, Cain and Abel. Cain became "a tiller of the ground," and Abel "a keeper of sheep," thus encapsulating the historic divergence of farmers and herders. And, of course, trouble soon followed to presage the conflicts that have marked events all down through the ages since humankind became settled and "civilized."

Cain, the sedentary farmer, became jealous because the deity "had respect unto Abel and unto his offering [of fat lambs]; but unto Cain and his offering [of the fruit of the ground] he had not respect."[107]

Subsequently Cain murdered Abel the nomadic shepherd and was himself cursed and driven from the land. He went away to "east of Eden" to become the first city builder in the "Land of Nod."[108]

This Biblical tale is surprisingly apt as a metaphorical mini-history of the rise of agriculture, for indeed it was not long after the first farmers that cities, those special harbingers of the arrival of "civilization,"[109] began to appear.

---

[105] The writers of the Bible would today be branded as male chauvinists. Of course, Eve was also expelled from Eden, and deservedly so.

[106] Holy Bible, King James Edition, Gen. 3:23.

[107] Ibid, Gen. 4:4-5

[108] "Nod" is a Hebrew word for "wandering." The Old Testament reports that Cain built a city and named it for his eldest son, Enoch, Gen. 4:17. Some archeologists believe that Uruk, a Sumerian city founded about 4000 B.C., may have inspired the reference in Genesis.

[109] The word "civilization" is derived through Old French from the Latin word "civis," meaning "citizen" and defined as an inhabitant of a city or town. Only

Agriculture as a Curse

Before we leave this subject, it's worth noting that according to the Old Testament account the deity was no big fan of farming. Not only did he favor the herdsman Abel over the farmer Cain, but he berated Adam: "In the sweat of thy face shalt thou eat bread, till thou return unto the ground; for out of it wast thou taken: for dust thou art, and unto dust shalt thou return."[110]

The deity, it seems, viewed a life of farming as a curse. That should not be surprising, since it was the crime of Adam and Eve to defy the natural world that was the deity's own achievement.

The roots of the Eden myth go back to prehistoric times long before the rise of Judaism, and remain a part of many other creation myths. It seems likely that the stories represent a kind of racial memory of the changes that took place as humanity assumed mastery over Nature.

Without that new-found power, cities were impossible things. Only when humans began to grow food from the ground instead of seeking it in the wild could first villages, then towns, and finally fortified centers emerge, wielding economic, religious and political power over thousands of inhabitants.

It may seem cynical to say so, but it was thanks to agriculture and due to the resulting rise of city-states that humanity discovered politics, slavery, warfare, and the Seven Deadly Sins[111] that have plagued our kind for thousands of years. Cities made it possible for some humans to become, like characters in George Orwell's *Animal*

---

with the rise of city-states could such a concept exist, since until that time all people were merely members of tribes or extended families.

[110] Holy Bible, King James Edition, Gen. 3:19.

[111] The Seven Deadly Sins were a creation of the Catholic Church. They included wrath, greed, sloth, pride, lust, envy and gluttony.

Farm, "more equal than others." Kings, the inheritors of Cain, ruled from ornate thrones in splendid palaces while masses of the descendants of Adam labored under the curse of agriculture.

Agriculture as the Parent of Cities

This pattern emerged wherever farming appeared. Agriculture was the parent of cities. Whether in the Levant, in the Indus Valley of southern Asia, the jungles of Central America, the lands along the River Nile, or the region between the Pacific Ocean and the Andes Mountains—in all these places and more the rise of agriculture was soon followed by the appearance of cities. Cities in turn gave birth to nations, generally through the "progress" of conquest.

There were several major centers of agricultural development. Those regions and the substance of their invention are:

• In the Near East between 9000 and 10,000 years ago, in the fabled "Fertile Crescent," in what is now Syria, Israel and Iraq. Here wheat, barley and other grains were joined by a variety of legumes, dates, olives and grapes. Domestication of sheep and goats were early innovations.

• In China, starting on the rich soil along the Yellow River at about the same time and expanding eastward. In 2853 BC, Emperor Sheng-Nung[112] of China named five sacred plants: soybeans, rice, wheat, barley, and millet. Cabbage and turnips are also products of China. Pigs were probably first domesticated there.

• In Central America, where farming and cities emerged between 9000 and 4000 years ago and spread north into what is

---

[112] Sheng-Nung was called "the Divine Farmer" and credited with the discovery of tea.

now the United States. Here corn (maize) was the keystone crop, joined by beans, peppers, squash, turkeys and dogs.

• In New Guinea, where a jungle-based agriculture may have emerged in the center of Papua-New Guinea as far back as 10,000 years ago. This was a garden form of agriculture centered on taro root, banana, sugarcane, and pigs.

• Along the western slopes of the Andes Mountains in modern day Peru and Ecuador about 6000 years ago, an agriculture based on more than 50 varieties of potatoes plus guinea pigs and llamas.

• In various parts of Africa, where coffee, sorghum, pearl millet, yams and oil palms were grown. Cattle were probably first domesticated in the Eastern Sahara, then a green and temperate place.

Perhaps the earliest stirrings of agriculture took place in the Levant, the region to the East of the Mediterranean Sea, and in particular in Mesopotamia, "the land between the rivers" (meaning the Tigris and Euphrates). It was here that Sumerian, Akkadian, Babylonian and Assyrian civilizations bloomed, faded, and disappeared each in turn. Their existence today in the region that is now Iraq is marked by numerous "tells," mounds built up, some over thousands of years, where ancient cities once stood.[113]

Mesopotamia may have been the model for the Garden of

---

[113] In his poem "Ozymandias," Percy Bysse Shelley evokes the fallen empires of ancient times. The poem describes a deserted statue of a once-great ruler, concluding:

"My name is Ozymandias, king of kings:
Look on my works, ye Mighty, and despair!"
Nothing beside remains. Round the decay
Of that colossal wreck, boundless and bare
The lone and level sands stretch far away."

Eden, and it was certainly the heart of the region that was historically known as the Fertile Crescent. Today it is a desert land, nearly every spot of fertile soil long since washed or blown away as a result of careless over-farming. The fall of these ancient civilizations should have been clear warning about the dangers of traveling the dead end path.

It is somewhat remarkable that the agricultural revolution began in many places at about the same time, soon after the end of the last Ice Age. Each burst was based on a major plant food, accompanied by an assembly of other plants and animals. In South Asia the foundation was rice; in Mexico, corn; in central Africa, grain sorghum; in Peru, the potato; and in northern China and Korea, millet. Except for the potato, all of these are hybrid grasses adapted to human use through selection.

For millennia humans had collected the seeds, fruits, and roots of many wild plants to supplement the wild game that was a prominent part of the hunter-gatherer diet. In the Fertile Crescent, the seeds of various wild grasses may have been collected for food as early as 15,000 or even 20,000 years ago.

In Mesopotamia the turning point that began the transformation of humans into farmers came soon after the end of the Younger Dryas, a 1400-year-long return to the Ice Ages, when a relatively unimportant grass evolved to create a new variety. An ancestral wild wheatgrass crossed with a type of goatgrass.[114] The result was a fertile hybrid that plant historians call emmer wheat.

Emmer had several important features. First, it grew larger seeds than its relatives, potentially yielding more food value, and thus attracting the attention of gatherers. Perhaps even more

---

[114] There are more than 20 varieties of *Aegilops*, or goatgrasses, found throughout the Middle East. They resemble winter wheat and are generally considered as a weed.

importantly, the seeds were attached to the surrounding husk in such a way that they were easily scattered in the wind. As a result, emmer quickly began to spread, soon becoming a dominant grass species.

The grass seeds were important as part of a hunter-gatherer economy because the kernels could be harvested and stored for use during lean times. As emmer became more widespread, it wasn't long before a new tool, the sickle, was invented to aid in harvesting the grain wherever it was found. Many Neolithic sickles were made of sharp bits of flint mounted on a curved stick or piece of horn, ideal to cut the stems for gathering. The grain was winnowed from the husks by hand with the help of the wind.

This was a step along the way to the arrival of farming, but the story didn't end there, for the harvest of emmer wheat was part of the natural world of gathering, not the product of agriculture.

In time the wild emmer crossed with yet another form of goatgrass to create a hybrid with even larger seeds. The plant was *Triticum aestivum*, bread wheat — but it had what in the normal flow of evolution would have been a fatal flaw. Its heads of grain were too tightly bound to blow apart in the wind, and could only fall to the ground where they grew. Unable to spread, the new grass variety would have been doomed to fail.

If there were, after all, a "Eureka" moment for agriculture, this may have been it. People who had formerly merely harvested grains of emmer wheat from wherever it grew now began to collect the seeds of bread wheat and do for them what Nature could not— move and plant them on fertile ground. That crucial step took them on the first tiny step to becoming farmers.

Our ancestors had emerged as *Homo agriculturis*. Nature began to be shouldered aside as humans took control of the environment.

Across the Fertile Crescent other plants were being domesticated, including einkorn wheat, hulled barley, peas, lentils, bitter vetch (a legume similar to lentils), chickpeas, and flax. Along with bread wheat, these eight crops joined to create the new model of food production in the Near East.

Like the gatherers, at about the same time hunters, too, were undergoing change. For millennia nomadic hunters had followed herds of wild sheep, goats and prehistoric cattle, acting as predators alongside the wolf and the lion. In time the hunters began to appoint themselves as protectors of the flocks, driving away the other predators.[115]

The domestication of animals was another side to the agricultural revolution, and went hand in hand with the development of food crops. It followed an orderly progression. First came the clever wolves, reshaped as dogs to serve as partners in hunting and herding. Another early adapter was the cat, which earned its way by protecting stores of food from rodents and other pests.[116] Next came the former wild prey itself, beginning with sheep and goats, tamed and adapted to use as sources of meat, wool, hides, and eventually, milk.

Another breakthrough came with the use of draft animals to aid in the hard work of farming. Onagers, a species of wild ass,

---

[115] The wolves were smarter than the lions, for they threw in their lot with humans who helped transform them into dogs, essential partners in hunting and herding. It should be kept in mind that today's many breeds of dogs are all of the same species as the wolf, transformed through selection by humans to perform various tasks but without evolving into a separate species. Theoretically, a gray wolf and a Chihuahua can breed and bear young.

[116] The cat is perhaps the only domesticated animal that was never part of a pack, flock or herd. Thus they have changed little from their wild ancestors and are well known for their independence and self-will. Perhaps due to their aloof and imperious nature, in ancient times cats were often held in deep respect and even worshipped as deities.

were first to be pressed into service, followed by oxen. These animals were put to work pulling wagons, carts and primitive plows, thus accelerating the ability of agriculture to support growing populations and their spreading cities.

About 5000 years ago the horse joined the extended biological family of agricultural humanity. Grander than any previous draft animal, horses were first used to pull chariots in battle and the processions of kings. Although the earliest domesticated horses were relatively small, breeding selection eventually resulted in larger breeds that were able to carry more weight.

Around 2000 B.C., somewhere in the Steppes of Central Asia, men[117] learned to ride astride the horse, thus assuming even greater mastery over Nature, not to mention their fellow men.

For thousands of years oxen and horses were associated with different social positions. Oxen could be fed more cheaply on poorer quality feedstuffs, so horses were luxuries for the relatively rich. The members of the top levels of society were horsemen, whether plowing their fields or riding into battle as cavalrymen or knights. Oxen, less glamorous, slow and ponderous, were the work animals of the peasant.[118]

The development of agriculture was a complex and diverse process, and from it grew all the blessings and curses of civilization. There were other elements, too, starting earlier than agriculture.

---

[117] There may have been female riders in ancient times, but there is scant evidence of it. Early horsemen were warriors and conquerors, and that was men's work. The myth of the centaur, half-man, half-horse, probably emerged from the era when riders first appeared out of Asia. Similar confusion was exhibited among the Aztec people of Mexico when the mounted knights of Hernán Cortez appeared in their midst, continuing a long tradition of conquest by hooves and the sword.

[118] J. G. Landels, "Engineering in the Ancient World," University of California Press, 1978.

The use of fire, perhaps the most important first step along the dead end path, took place even well before modern man came on the scene. As a part of Nature, fire did not need to be discovered, but learning to control it was an important event in pre-history. It appears that fire may have been used by the primitive hominids known as Homo erectus as early as 1.8 million years ago, but until recently it was believed they could not start it but used fire ignited by lightning or other natural causes. However, a 2008 archeological dig in Israel revealed evidence that the early human relatives were making fire as long as 790,000 years ago by striking flints on stones containing iron.[119]

Prehistoric Chefs and Brew Masters

Following the taming of fire came the invention of cooking. For thousands of years most cooking may have consisted merely of roasting meat or tubers in a fire, making them more appetite pleasing and easier to digest.

It has been suggested that cooking food made more energy available for our ancestors, allowing them to devote more energy to the development of larger brains.

For example, in a recent book[120] Richard Wrangham, a primatologist and anthropologist, reported on four decades of observing chimpanzee eating patterns. In an interview with *The New York Times*, Wrangham said:

> ...our large brain and the shape of our bodies are the product (sic) of a rich diet that was only available to us after we began cooking our foods. It was cooking that

---

[119] "Fire Out of Africa," Press Release, 27 October, 2008, Hebrew University of Jerusalem.

[120] Richard Wrangham, "Catching Fire: How Cooking Made Us Human," Basic Books, 2009.

provided our bodies with more energy than we'd previously obtained as foraging animals eating raw food."[121]

The Harvard scientist theorizes that it was through the adoption of meat eating that Homo habilis, a primitive hominid, evolved into Homo erectus about 1.8 million years ago. That was a giant evolutionary step toward the emergence of modern humans. As we have seen, members of *Homo erectus* were perhaps the first short order cooks, possibly contributing to their evolution into present-day humans.

Much later, in the millennia leading up to the emergence of agriculture, cooking played an essential role as gatherers began to collect grass seeds for food. The seeds are essentially inedible without being processed and baked or boiled. Gatherers ground or pounded the grains to break up the seeds and reduce them to flour. This was generally used to make flat breads or cakes,[122] most likely cooked on hot stones. It was these primitive breadstuffs made from coarse, whole grain flour that came to be called "the staff of life."

Early breads made from emmer, bread wheat, or barley were probably far more nutritious than most of what we encounter today. They were baked from whole grains instead of "enriched" white wheat flour that has been separated from the germ and turned into a bland and dismal product. No longer nutritious foods in their own right, atrocities made of "enriched" white flour and with names like "Wonder" and "Rainbo" are hardly treated as foods for their own sake, but merely as a convenient base on which various other substances are smeared or stacked (think PB&J, BLT, cheese or baloney sandwiches).

---

[121] "From Studying Chimps, a Theory on Cooking," New York Times, April 21, 2009.

[122] Many forms of flat breads and cakes are still used around the world, including pita, matzo, nan, tortillas, pancakes and others.

Another early use for grain, one that was sure to have gained quick popularity, was the invention of brewing. Like many discoveries, the "invention" may have occurred by accident when someone decided to eat "spoiled" fruit or berries that had fermented and found that he wanted more. No doubt the advent of the neighborhood tavern was an early feature of the new cities, along with the associated problems of alcoholism that have continued to plague society to this day.

Archeologists have found very early evidence of mead-making in China and by about 8000 years ago beer was being brewed from emmer wheat at Ur, the mythical home of Abraham and the prototypical city-state of the Levant. The Sumerians, first to bring civilization to the Middle East, credited their fertility goddess Nin-Harra as the inventor of beer. Early brewing sites have been discovered in northern Iran and other places.

At least one researcher believes that the making of alcoholic beverages was a prime motivator in the transition to agriculture. In a recent book[123] archeologist Patrick McGovern of the University of Pennsylvania points out that early bread-making did not yield a very palatable result and that farmers may have preferred to use the grain and other foods they grew to make beer, wine and other alcoholic drinks.

In an interview published in the German magazine *Spiegel*, McGovern stated: "Available evidence suggests that our ancestors in Asia, Mexico, and Africa cultivated wheat, rice, corn, barley, and millet primarily for the purpose of producing alcoholic beverages."[124] He speculates that moderate alcohol consumption

---

[123] Patrick E. McGovern, "Uncorking the Past: The Quest for Wine, Beer, and Other Alcoholic Beverages," University of California Press, October, 2009. McGovern is an expert on identifying traces of alcohol production from early sites.
[124] "Brewing Up a Civilization," Spiegel Online, 24 December, 2009.

may have been a benefit to early Neolithic imbibers, providing concentrated carbohydrate energy. (Distilled spirits, of course, came much later.)

Winemaking was another early adaptation to the making of alcoholic beverages, leading to the domestication of the grape and development of viniculture as a sub-set of agriculture, particularly in the Levant and southern European regions where sun and soil were well adapted to the cultivation of vines. Similarly, the art of beekeeping may have begun for honey's value as the main ingredient in mead.

Technologic Advances Increased Production

As the years passed various technological advances accelerated not only the spread of agriculture, but also the intensity of its production. One such was the scythe-and-cradle, a tool that allowed a reaper to harvest ripe grain faster and more efficiently than with a hand sickle. As the Late Neolithic passed into the Bronze Age, metal versions of the earliest tillage tools appeared— the spade, mattock, pick, fork and rake, all derived from the most basic digging sticks and stone hoes.

But those were minor advances beside the wheel, considered as one of the greatest inventions of history. It appeared in the Near East about 7000 years ago. Farmers enjoyed the convenience of four-wheeled carts to carry their harvest, and warriors rode two-wheeled chariots to defend their city-states from invaders, or to conquer new lands for their rulers.

The wheel was important, but another innovation may have had the greatest effect of all. That was the plow. Almost the very symbol of agriculture, it was a device for ripping up the land. Prior to that fields could be prepared only with digging sticks and crude hoes, hard, back-breaking work that limited the amount of land that could be farmed by a given number of people and how deeply they could till the soil.

The earliest plows appeared in Mesopotamia about 5000 years ago. At first they were crude affairs called "scratch plows," made of wood and drawn by oxen. They simply gouged a line in the soil into which seeds could be planted. Later, horses were put into harness to draw plows of increasing sophistication. So-called moldboard plows turned the ground in furrows, aerating the soil and bringing up nutrients from below.

The coming of the Iron Age caused further improvements. The Celts, for example, were master blacksmiths who as early as 400 BC forged plowshares and made iron rims for the wheels on their wagons and carts. Romans engineered the colter, a device to slice the ground ahead of the plowshare, making it easier to turn the soil.

Many years were to pass before further major improvements would be made in agricultural technology. It has been said that the Romans had better plows than were available to George Washington nearly two millennia later.

In truth, there was little difference between the tools and methods of a Roman plowman at the time of Christ and my grandfather in 1944, each steering a plow behind a single horse or team.

Mastery of water was another key element in the advance of agriculture, for when crops could be irrigated during dry spells a successful harvest could be assured. In Mesopotamia and Egypt the art of irrigation reached a high peak, as shown by the many water channels, dikes and holding ponds that are revealed at archeo-logical sites. Pumping of water began early in pre-historic agriculture, beginning with a simple device variously called the swing-beam or swipe. Basically a counter-balanced lever used to dip a bucket into a well or canal and raise it to pour into a channel or receptacle. Although labor-intensive, these are still seen in some less developed areas of the world.

When water needed to be lifted only a short distance, the screw pump was often used. This was a tube with a screw-like insert. One end would be placed in the water and as the screw was turned it would draw water up through the tube and out the top end where it could flow away to fill a tank or water a field. Another type was the drum pump, which worked kind of like a water wheel in reverse. As the wheel turned it lifted water into a channel.

The "Second Agricultural Revolution"

Once again, except for slow and incremental improvements, tools such as these were used right up to the beginning of the industrial revolution. It was only when fossil fuels were put to work to replace human and animal power that industrial agriculture put us on a fast track down the dead end path. To a great extent, what we call the Industrial Revolution might better be termed the Second Agricultural Revolution.

There are many connections between agriculture and the natural world. In the succeeding chapters we'll take a look at the inputs on which agriculture depends, particularly in its latest industrial form. For simplicity, I've divided the subject into four parts, taking a cue from the Greek philosopher Aristotle who believed in four elements from which everything else is made. He called these elements Earth, Air, Fire and Water.

In my usage, these four elements stand for the basic requirements of farming—Earth for the soil; Air for sunlight and the climate; Fire for energy; and Water for, well, for itself. We'll examine each of these subjects in the next four chapters.

David L. Brown

Chapter Six

The First Element: Earth

What is the soil? Is it merely the dirt beneath our feet, inconvenient mud on our shoes when it rains, dust blowing in our faces when it doesn't? For many people today, that may be all they know or care about the soil that is the source of all land life on our planet.

Soil is not just "dirt," but a living, breathing, evolving thing. The ecologist James Lovelock[125] has written about the Earth as a metaphoric living being he calls "Gaia." This is a powerful concept, and should you think about it you would find the soil at its very heart. Indeed, if there were no soil there could be virtually no land life on the planet.

Fertile, healthy soil is a precious and complex substance, forming an oh-so-thin and fragile layer on the surface of the Earth. On average it is only about six inches thick, more in some places, non-existent in others. It consists of a matrix of minerals, organic

---

[125] Lovelock has written six books on his Gaia theory, most recently "The Vanishing Face of Gaia: A Final Warning," Basic Books, April, 2009. Born July 26, 1919, in his nineties Lovelock is scheduled to fly into space from Spaceport America. The flight will be a gift from Sir Richard Branson the creator of the civilian space enterprise now being built in New Mexico.

matter, and life forms such as bacteria, nematodes, and earth-worms, all working together to create a rich environment from which plants can thrive. It is a slow and steady process.

In a natural state, it's within the soil that a cycle of continual renewal takes place, defining a never-ending circle between plant and animal life. In this cycle, plants absorb CO_2, grow, emit oxygen, and are consumed by animals. Members of the animal kingdom in turn breathe oxygen, emit CO_2, and return nutrients to the soil as urine, feces, and ultimately their bodies.

Between Plant and Animal Kingdoms

The soil is the intermediary between the two realms of animals and plants. Below the surface of the living soil organic materials and minerals are broken down, processed, and incorporated into forms that new plant growth can use. The cycle continues, as it has for hundreds of millions of years ... until the arrival of *Homo agriculturis* and our meddling ways.

There's a stiff price to pay for over-farming the soil, particularly through the use of artificial fertilizer. All too often the unnatural substances become pollutants, leaching into water supplies, flowing down rivers and streams, and creating ocean "dead zones." They even contribute to greenhouse gas accumulation in the atmosphere.

Here is a brief overview of the situation according to the authors of a recent paper that appeared in *Science* magazine:

> "Harvested crops remove nitrogen, phosphorus, and other nutrients from agricultural soils—and sustaining agricultural production requires replacing those nutrients, whether through biological processes like nitrogen fixation or through the addition of animal wastes or mineral fertilizer to fields. Globally, fertilizer is

the major pathway of nutrient addition; it has more than doubled the quantities of new nitrogen and phosphorus entering the terrestrial biosphere. These inputs have helped to keep world crop productivity ahead of human population growth and can enhance rural economic development. However, environmental costs of nutrient pollution from agriculture have been substantial, including the degradation of downstream water quality and eutrophication of coastal marine ecosystems, the development of photochemical smog, and rising global concentrations of the powerful greenhouse gas nitrous oxide."[126]

There are major disconnections between the natural cycle and modern agriculture, even when the latter is practiced with what are considered sound methods. Like oil and water, industrial farming and Nature just don't mix.

I have seen a diagram[127] showing the way in which a university extension service believes the cycle between the animal and plant worlds is supposed to work on a well-managed modern farm.

Let's analyze this scenario to see how it fits into long-term sustainability and the protection of the soil. The flow chart is supposed to suggest how a farmer can manage his soil to maintain it in balance. Before we start our analysis, note that the extension service diagram shows that:

- Animals are removed from the soil (no pasture is used).
- Some manure is returned to the land using a manure spreader, but nitrogen in the form

[126] P.M Vitousek et al., "Nutrient Imbalances in Agricultural Development," Science, 19 June, 2009.

[127] From "Nutrient Cycling & Maintaining Soil Fertility," undated bulletin, The Ohio State University Extension.

of ammonia (NH_3) escapes into the air.
- At least some of the crop is sold off the farm.
- Some crop residues are returned to the soil.
- Nitrogen fertilizer is added.
- Runoff losses are indicated.

In all-out industrial agriculture as presently practiced in America, the odds are that on most farms none of these actually occur to any meaningful extent ... except of course for the runoff of topsoil. Consider the following rebuttals to the points listed above:

- First, in many cases animals have not only been removed from the land but taken entirely off the farm and grown elsewhere in factory-style confinement pens or barns. Many farms, and especially the larger operations that have become common, are no longer diversified, but dedicated to producing as few as one or two kinds of crops, typically corn and soybeans in the Midwest, for example. This trend to specialized mono-crop cultivation is more the norm than the exception. Diversified family farms are becoming relics of the past, replaced by giant operations trapped in the need for "progress" to create profits.

- If animals are completely removed from the farm, their excretions not only are lost to the soil but end up elsewhere as industrial waste products and potential pollutants.

- Crops that are sold off of the farm are also consumed else-where and those resulting wastes, whether animal or human, also are lost to the soil from which they grew.

- Nitrogen is the most important substance for plant growth, but it is far from the only essential ingredient in healthy soil (and in fact the injection of toxic anhydrous ammonia is certainly not advantageous to the essential biological residents of the soil).

• Crop residues often are not returned to the soil either, especially when fields are harvested as hay or silage that's sold off the farm. Plans to use crop residues or to grow and sell crops such as switch grass for the production of ethanol promise to completely separate almost all crop residues from the land. Organic matter is essential for the maintenance of healthy topsoil. Under sustained industrial farming the relationship has become one of taking from the soil while giving back nothing but unnatural chemicals. This amounts to the deliberate depletion of a critical natural resource.

• Thus, in this scenario, the only certainty is the loss of topsoil due to runoff.

Industrial Farming Starves the Soil

Only in a completely closed system such as Nature originally devised can the soil be maintained in top condition … and that, of course, is impossible to achieve through the large-scale, highly specialized technological farming that's become the norm in the United States.

It would be difficult enough to maintain ecological balance even on very small, family-sized farms where most of the produce is consumed on-site, because at least some losses will likely be suffered. To attain long-term sustainability, such a small-scale farm enterprise would need to replace at least the nutrients that are removed from the topsoil, and ideally add even more to help improve the soil. It's unlikely that could be achieved through the use of artificial inputs, especially as fossil fuels and other natural resources become scarce and expensive. It seems likely that anything short of a complete cycle reflecting the original "intent" and methods of Nature is starving the soil.

That process began as soon as humans became farmers, but until the start of the industrial age with its mechanization and use of fossil fuels the process was relatively benign as long as respon—sible forms of agriculture were practiced. Farms were smaller, farmers were closer to their land, and many took seriously their responsibilities as stewards of the soil.

There are still many farmers who try to maintain the richness of the soil, but most of today's food production comes from all-out industrial farming. Wherever technological agriculture and factory farming are practiced, they not only break the cycle of Nature, they utterly destroy it. As the process unfolds the soil sickens and dies a slow death. That's been happening at an accelerating rate ever since fossil fuel driven technology has been applied to usurp Nature.

It doesn't take much thought to see the problem with that. It's as if Nature long ago provided us with a very large cookie jar, filled to the brim. Our ancestors began to take a cookie from time to time, then some more, and finally in our present age we're grabbing them by the handful. But the cookie jar is only so big, there are only so many cookies, so what is going to happen next?

As we saw in Chapter Two, human beings and our extended family of domesticated animals comprise as much as 98 percent of the total mass of land animals on Earth.

To a large extent, and particularly in the "advanced" nations where industrial farming is the norm, the output of excretions and the deceased from this mass of humans and animals are seques-tered. The debt to the soil goes unpaid.

Many of our organic waste products are buried in landfills, processed through sewage plants, or allowed to flow down rivers into the sea. That applies not only to human waste but in many cases to that generated by livestock grown in feedlots and on factory farms. Their manure and urine are treated as unwanted pollutants rather then as the precious substances they are.

When we die, our bodies are preserved and buried in sealed caskets or burned and the sterile ashes stored in urns or scattered in streams and rivers. We have removed even our own bodies from the cycle of Nature and thus the soil is denied payment of our last debt to Gaia. What should be honored as a final return to the Earth that sustained us has been twisted and broken.

Damaged Soils are Susceptible to Erosion

Slow nutrient loss is only part of the problem. Soils that are dead or dying are subject to erosion by wind and water. Rich loam, left to grow continuous perennial grasses or legumes, will sustain and improve itself over time. It will hold moisture for use by growing plant life, and welcome the touch of the rain.

Not only are our soils being starved to death, they are eroding through carelessness and mismanagement.

When soil is deprived of its organic material, when its bacteria and earthworms are gone, nothing remains but a mineral residue. Vulnerable to wind and water, the dead remains of once-fertile soil blow and wash away leaving only sterile sand, rocks and clay, a desert landscape.

Soil loss is a serious concern, as Cornell University ecologist David Pimentel has discovered. In a 2006 news release[128] announcing the results of his study of soil loss, he was quoted:

> Soil erosion is second only to population growth as the biggest environmental problem the world faces. Yet, the problem, which is growing ever more critical, is being ignored because who gets excited about dirt?

---

[128] "Slow, insidious soil erosion threatens human health and welfare as well as the environment, Cornell study asserts," news release, Cornell University, March 20, 2006. Pimental's paper appeared in the Journal of the Environment, Development and Sustainability, Vol. 8, 2006.

Indeed, few of us even think about the soil, although according to Pimentel 99.7 percent of all human food comes directly or indirectly from cropland. Ominously, he reported that the amount of arable land in the world is shrinking by nearly 37,000 square miles each year due to erosion. That's an area the size of the state of Indiana, lost forever to humanity.

A serious problem indeed, and one that needs equally serious attention if we are to avert a coming famine—for without soil there will be no food, and without food ... well, you finish the thought. At the time he published his study Pimentel estimated that 3.7 billion humans were already malnourished. These are the "ben–eficiaries" of the Green Revolution.[129]

Erosion may not seem to be important to most people because they tend to look at the short-term. One rainstorm can easily cause a vulnerable field to lose one millimeter of soil (about 1/25 of an inch). It doesn't seem like much, perhaps even insignificant—but that's a false impression. That silly little millimeter is equal to about five tons of topsoil per acre, according to Pimental.

And if that doesn't grab your attention, consider this: If left to herself, it takes Nature about 500 years to produce a single inch of fertile topsoil. If losses proceed at the rate of one millimeter each time a significant rainfall occurs, in just 25 such events an amount of soil will have been lost that would take natural processes half a millennium to replace. In the Midwestern United States it's easy to imagine 25 such rains within just a few seasons.

Grim as it may seem, in some places this is a best-case scenario. Here are some additional factoids from Pimental's survey,

---

[129] Note that this figure is considerably higher than the estimate of "more than a billion" put forth by the UN, as quoted in Chapter One. It seems that how we define malnourishment—where do we draw the line—can yield different results. If we count anyone who goes to bed hungry, the number can be quite large, as compared to the result if we consider only the actual cases of starvation.

which drew data from more than 125 sources around the world. Here, with some side comments in parentheses, are some of his discoveries:

• The United States is losing soil ten times faster, and India and China 30 times faster, than the natural rate of replacement. (This could be the very definition of "unsustainable.")

• In the United States, soil erosion costs about $37.6 billion each year. Worldwide, the cost of soil erosion is around $400 billion annually. (These are direct costs, not considering the long-term effects of environmental damage.)

• In the 40 years from 1966 to 2006, 30 percent of the world's arable land has become unproductive. (During that same period world population doubled.)

• About 60 percent of the soil that washes away ends up in rivers, streams and lakes, leading to more serious flooding and contamination with fertilizer and pesticides. (Troubles with water resources are a key concern, and will be addressed in a later chapter.)

Concern about erosion is nothing new. Writing nearly three decades ago, Neil Sampson, an executive with the National Association of Conservation Districts, wrote a book in which he noted that "[t]here are places in America where the entire layer of topsoil has been lost in only 50-100 years of cultivation."[130] He added:

> "Protecting the quality of our nation's topsoil—and therefore much of the quality of our lives—is largely within human control. Good management can improve

[130] R. Neil Sampson, "Farmland or Wasteland: A Time to Choose," Rodale Press, 1981.

topsoil quality, but bad management can destroy it within a few years. A cropping system that takes out more nutrients and organic matter than it replaces is essentially "mining" the land, making it susceptible to the main agent of destruction: accelerated erosion. Topsoil lost in this manner is the result of human neglect and greed."

Since Samson addressed this subject, the intensity of industrial farming has increased, with more fertilizer, more chemicals, higher-yielding seeds, more aggressive tillage, factory farming—all draining the natural goodness from the soil and setting it up for erosion.

As we saw in the last chapter, the depletion of soil has been the bane of farmers for millennia. Many parts of the formerly "Fertile Crescent" lands of Mesopotamia have turned to desert today, the result of centuries of neglect. In the early 1950s conservationist Walter C. Lowdermilk issued a report[131] on the history of that region as told by its soil. He concluded that at least eleven empires had risen and fallen there over several thousand years.

In Syria he found the ruins of more than a hundred dead villages and market towns, many no doubt still viable at the time of Christ. He recounts how some of the ruins of those towns are not buried—they stand atop bare rock, all the soil and even the sand and dust having been eroded away for miles in every direction. He wrote:

"If the soils had remained, even though the cities were destroyed and the populations dispersed, the area might have been re-peopled again and the cities rebuilt. But now that the soils are gone, all is gone."

---

[131] Walter C. Lowdermilk, "Conquest of the land through 7000 years," Agricultural Information Bulletin, U.S. Dept of Agriculture, Soil Conservation Service, 1953.

The devastation in that area has continued to this day, leading to predictions that the entire region is probably doomed to desertification, creating a new Sahara. In 2009 an article[132] *in New Scientist* magazine stated:

> "With the region beset by drought and a slew of pro-jected new dams in the pipeline, it is looking increasingly likely that the Mesopotamian cradle of civilization will become a desert.

> "In ancient times the valleys of the Tigris and Euphrates rivers through Iraq were bountiful, sustaining civilizations such as Sumer and cities such as Babylon. That stands in stark contrast to a detailed assessment of the region's future under climate change, published in 2007 by Japanese and Israeli meteorologists. This suggested flow on the Euphrates could fall by 73 percent, with the authors warning that the ongoing drought in the region was likely to become permanent."

The article quoted Akio Kitoh, lead author of the report[133]: "The ancient Fertile Crescent will disappear in this century. The process has already begun."

Soil loss has long been recognized, but until the early part of the last century it wasn't seriously addressed in America because there was always lots of new land to the West that could be homesteaded. To a large extent westward expansion was driven by pioneers who had "farmed out" their former land.

---

[132] Fred Pearce, "Desert future for land that once nourished Babylon," New Scientist, 1 August, 2009.

[133] A. Kitoh, A. Yatagai, P. Albert, "First super-high-resolution model projection that the ancient 'Fertile Crescent' will disappear in this century," Hydrological Research Letters, January 17, 2008.

By the 1920s the westward expansion was coming to an end—there was no more arable land to be found—and it began to dawn on some that soil loss was an emerging disaster. A young soil scientist from North Carolina, Hugh H. Bennett, began to speak and write on the dangers of erosion. He was convinced that the destruction of soil was causing serious harm to the nation. Eventually he persuaded the U.S. Department of Agriculture (USDA) to publish a paper on the subject.[134]

That got the attention of Congressman James P. Buchanan of Texas, who called hearings to address the subject. In 1929 Congress voted a grant of $160,000 for the USDA to investigate "the causes of soil erosion and the possibility of increasing the absorption of rainfall by the soil in the United States."

Soil advocate Bennett continued to lobby, leading to a program to carry out demonstration projects to show farmers how erosion could be prevented. He discovered that gaining public attention to the problem was a challenge, but nature stepped in with a helping hand. In 1934, just as the first survey on soil quality and erosion was being completed, something new came to America: dust storms.

Millions of acres of semi-arid soil in the upper Great Plains had been opened to cropping, heeding a mistaken idea that "rain will follow the plow." Beginning in 1891, William Smythe, a self-proclaimed prophet who believed that deserts could be made green, began to promote "the somewhat metaphysical nineteenth-century notion that the very acts of tilling and planting might increase precipitation that would in turn sustain more farming."[135] Smythe's ideas were based less on science than upon his readings of

---

[134] *Soil Erosion: A National Menace,* USDA Circular 33, 1928.

[135] P.S. Kibel and P.D. Batchelder, "Rain Follows the Plow: An Introduction to the Issue," Symposium Paper, Golden Gate School of Law, October 4, 2009.

Biblical texts, but it encouraged tens of thousands of Americans to flock to the semi-arid lands of the Great Plains.

By the early 1930s the promised rain had failed to appear. Drought caused the disturbed soil to dry out and turn to dust. On May 12, 1934, a devastating windstorm hit the region, sweeping up the dusty soil and blowing millions of tons of it far away. Looking back several years later, Bennett recognized the event as a turning point:

> "This particular dust storm blotted out the sun over the nation's capital, drove grit between the teeth of New Yorkers, and scattered dust on the decks of ships 200 miles out to sea. I suspect that when people along the seaboard of the eastern United States began to taste fresh soil from the plains 2,000 miles away, many of them realized for the first time that somewhere something had gone wrong with the land."[136]

The "Dust Bowl" got its name in the following year after an even worse storm on April 14, 1935, a date that was dubbed Black Sunday. More and more dust storms had been blowing up in the years leading up to that day. They were first seen in 1932, when 14 were recorded on the Plains. In 1933, there were 38, and by 1934 an estimated 100 million acres of former prairie land had lost all or most of their topsoil due to wind erosion. The storms drove hundreds of thousands of settlers off the land. Many, as chronicled by novelist John Steinbeck in his novel *The Grapes of Wrath*, migrated to California.[137]

---

[136] Hugh H. Bennett, "The Land We Defend," a speech given in Milwaukee, WI, July 2, 1940.

[137] I recall a conversation in about 1978 with an old-time ranch woman who had lived all her life along the Musselshell River in east-central Montana. As we had coffee together, she pointed out of her kitchen window to the South and told me

The Dust Bowl conditions at last focused public attention. On March 25, 1935, the Soil Conservation Service[138] became a permanent agency of the USDA. Bennett was named to head it.

But as legislators should have learned time and again, it's one thing to pass laws, and quite another to change reality. The problem of soil erosion didn't go away. In 1939 Bennett appeared before Congress with discouraging news. He said:

> "In the short life of this country we have essentially destroyed 282,000,000 acres of land, cropland and rangeland. Erosion is destructively active on 775,000,000 additional acres. About 100,000,000 acres of cropland, much of it representing the best cropland that we have, is finished in this country. We cannot restore it... We are losing every day as the result of erosion the equivalent of 200 40-acre farms."[139]

President Roosevelt put the situation into perspective with his usual sharply focused words. In a 1937 letter to the 48 governors urging action at the state level, he wrote: "The nation that destroys its soil destroys itself."

And yet despite all that, nearly fifty years later the situation had not only failed to improve, it had grown worse. In 1980 Agriculture Secretary Bob Bergland estimated that more than ten percent of the land being farmed in the U.S. was eroding at more than 14 tons per acre each year. "If something is not done about that land, it will be barren," he said in reaction to an alarming study that showed that American farms were losing topsoil at the fastest rate in history, faster even than during the Dust Bowl era.

---

that when she was a young woman the distant landscape was dotted each night with the lights of homesteader's cabins. None survived the 1930s.

[138] The agency is now the Natural Resources Conservation Service.

[139] Quoted by Peter R. Huesay of The Environmental Fund, in 1977.

The problem had accelerated in 1972 when the Soviet Union suffered a disastrous crop failure and suddenly announced it would buy 30 million tons of American grain. Russia had previously been a breadbasket nation, and under Soviet rule vast tracts of land were opened to farming in an effort for the nation to remain self-sufficient. Peasants were assigned tractors to become industrial farmers ... and as in the High Plains, the effort was a miserable failure.

Earl Butz, then Secretary of Agriculture under Richard Nixon, helped arrange the sale of grain in the hope of giving a boost to corn prices.[140] Butz was a prime advocate of industrial farming, urging farmers to "get big or get out," and to plant their land "fencerow to fencerow." For a few years farmers (and agribusinesses) enjoyed a boom, but as with all bubbles, bust was soon to follow. The resulting agricultural depression caused thousands of farmers to fail, and the damage to the soil from overly aggressive farming was significant.

I have a photo from the mid-1970s of myself with Sec. Butz at a Washington reception for agricultural editors. At that time I was part of the problem, working as an advocate for industrial agriculture.

More recently some of the more fragile land has been taken out of crop production and soil loss has dropped, but as reported above from Cornell ecologist David Pimental, erosion continues to be a serious problem today.

Seldom is the effect of erosion more clearly illustrated than by an anecdote told me by a friend. Around 1975 he visited a farm

---

[140] Butz's actions were said to be politically motivated, aimed at taking farm votes away from George McGovern, the Democratic presidential candidate in 1972. Nixon won re-election by a landslide, with McGovern carrying only Massachusetts and the District of Columbia.

near Triplett, MO that belonged to an acquaintance of his father. The farm had been in the same family since the sod was first broken. For some reason a single acre was left untouched down through the years, a kind of souvenir of the past I suppose. As my friend reports, that single acre of preserved prairie had become a raised platform, or as he said, a "plateau," standing a foot or more above the surrounding land that had been plowed and cropped for about a hundred years. "It was quite startling," he recalled.[141]

Just as present deniers are casting doubt on the reality of global warming and climate change (something that has been clearly accepted by the vast majority of scientists), there have been voices to tell the public that soil erosion is no problem. Here's an almost amusing example. You may recall the late Julian L. Simon, whom we met in Chapter Two. He was formerly an economist with the University of Illinois and later joined the Cato Institute, a Libertarian think tank.

In 1980 Simon wrote in *Science* magazine[142] that the problem was not a shortage of land, but of people. His ideas were widely discussed, creating confusion among the general public. His "theory," if it deserves that name, was that the worldwide supply of arable land had actually been increasing; that food was becoming cheaper and more abundant; and that the environment in the United States had been improving.

He offered a computer model[143] purporting to show that as population increases, so does productivity, concluding that the best thing that could happen to America would be to add more and

---

[141] Private communication, Val Germann, Columbia, MO.

[142] Julian L. Simon, "Resources, Population, Environment: An Over-supply of False Bad News," *Science*, June 27, 1980, p. 1,432.

[143] Mark Twain's famous statement, "Figures don't lie, but liars figure," is truer than ever in our computer age. It is perhaps possible to devise a computer program to "prove" absolutely anything, no matter how ridiculous.

more people, thus assuring the advance of invention, technology and knowledge.

Along with a few facts, common sense alone seems adequate to counter Simon's ideas. Like many other economists, he conveniently ignored the fact that land and fertile soil are limited resources and subject to decline. In later chapters we'll examine at length the role of economists in the advancement of industrial agriculture.

Beyond the broken cycle of nature and resulting erosion, there are other threats to the soil. As we saw in the last chapter, so-called minimum tillage or no-till farming is no friend of natural topsoil. Not only is the land saturated with ag chemicals, many of them toxic, but the soil is repeatedly compressed beneath the weight of giant tractors and combine harvesters. The soil eventually develops a "hardpan" layer that prevents plant roots from penetrating into the subsoil. In a natural state, and with the aid of soil biota, those roots would grow deep, thus aerating and helping enrich subsoil and helping it along the way to becoming topsoil for the nurturing of future crops.

Another ominous trend in places such as Brazil, Congo and Indonesia is the destruction of rain forests for purposes of growing crops. The amount of topsoil in these tropical forests is usually thin and fragile, and it's not unusual for it to be "farmed out" in a decade or less. The remaining soil turns into a clay-like crust, and the farmers move on to cut even more forest trees to grow soybeans, sugar cane or some other unnatural monocrop for another few years. They leave behind conditions similar to those in present-day Iraq—but in devastated rain forest lands, the processes that took millennia in the Middle East can run their course in a matter of a few years.

There's another way in which cropland is disappearing in

America and elsewhere, and that's through the steady encroach-
ment of human development. Each year tens of thousands of acres
disappear beneath roads, houses, airports, parking lots, schools,
shopping centers and all the other depredations of cities as they
expand in response to growing populations. Once the topsoil is
removed and concrete or asphalt laid down, what are the chances
that the land beneath will again become productive on any time
scale that matters to human society?

It's ironic that although agriculture is parent to the city, cities
themselves have a way of eating up the very land and other
resources on which agriculture depends. It's reminiscent of the
mythical Ouroboros, a snake that swallowed its own tail.

In the same way that cities are devouring the land around
them, industrial agriculture is eating up the rest, the basis for all life
on land. Through careful management and non-industrial
methods, it's possible to maintain and even improve the soil.
Instead we are for the most part squandering it as if there would
always be more. This is an act of tragic hubris, of careless greed for
which future humans will hold us culpable throughout all coming
times.

Chapter Seven

The Second Element: Air

One can hardly open a magazine, newspaper or internet news page or turn on a radio or TV without learning some dire new fact about climate change, global warming, air pollution, drought, blowing dust, floods, and other "unnatural disasters." All of these relate to Air, our second "element" of agriculture. If the soil is the heart of Gaia, the atmosphere is her lungs. Just as *Homo agriculturis* has disturbed the Earth below, so we have harmed the dominions of the air.

What comes from above is essential to the natural cycle of Nature. From the sky come sunlight and rain. The air is the reservoir of carbon dioxide and oxygen shared back and forth between plants and animals.

The atmosphere allows the Sun's rays to shine through, providing energy that plants convert into growth through photosynthesis. It filters out harmful ultraviolet rays that would make life impossible.

The air is the mediator of climate, a complex and inter-connected system that rules the weather and tempers heat. Through the wind, the atmosphere distributes seeds and pollen, allows birds and insects to fly, and generally makes the Earth work.

This is not a book about global warming and resulting climate change[144]—but they are important parts of the story that is now unfolding. Humans have poured rising amounts of greenhouse gases into the atmosphere by burning carboniferous fossil fuels that took millions of years to be sequestered in the Earth. This is one of the important ways in which we have disturbed the balance of Nature through industrialization and in particular technological farming.

I assume that most readers of this book are fairly well informed about global warming and climate change. There has been a lot of noise about the subject in recent years. Perhaps you watched the movie by Al Gore, or read interviews with scientists such as James Hansen. Unfortunately you probably have also heard loud claims that global warming isn't real, that it is a "hoax" or fraud.[145] Almost everyone it seems has jumped on this subject, including columnists, politicians, evangelists, lobbyists, bloggers and business-men—the vast majority of whom, no matter which side they're on, are grossly uninformed, biased, or both.

Sadly those least qualified have made the most noise. In this atmosphere of disinformation, it's difficult for members of the general public to realize that the vast majority of scientists who are trained in relevant disciplines agree that global warming is real, and that it poses a serious threat, although they are not in agree-

---

[144] Anthropogenic global warming and climate change are really two faces of the same thing, linked by cause-and-effect. No understanding of the one can be reached without consideration of the other. "Anthropogenic" means "human caused." It should be pointed out that "climate" and "weather" are quite different things and are often confused.

[145] Although it's easy to see who might benefit from denial of global warming (hints: Saudi Arabia, ExxonMobil, et al.), it's difficult to imagine why thousands of scientists would engage in an enormous hoax. What's in it for them?

ment on just how much danger we face. The more ominous predictions have been lowered in recent years, but there is no doubt that climate change is taking place.

And again, to a great extent it's agriculture, and in particular its present technological form based on resource depletion, that is the most serious problem facing humanity. The warming of the planet and climate change are only secondary effects of industrial "development."

Here's the chain of evidence linking agriculture with climate change:

AGRICULTURE > POPULATION GROWTH > INCREASED USE OF

FOSSIL FUELS > GLOBAL WARMING > CLIMATE CHANGE

All of these effects link directly back to the beginning of agriculture, that "original sin" against nature. It's worth noting that each step in this series of links could be diagrammed as feedback cycles. That is, the more population growth, the more aggressive agriculture must become. The more fossil fuels are used to increase food production, the more population can expand. And, of course, the more global warming that occurs, the more the Earth's climate is likely to change, and vice versa.

This is a toxic mix of causes and effects. Since the Industrial Revolution global warming and climate change have been building like a witch's brew to become the greatest threat facing humanity.

But forget global warming and climate change. Yes, you read that right. The real problem is industrial agriculture. Global warming and climate change are merely symptoms. If humans fail to control industrial agriculture, Nature will take care of all of the problems that face us, and not in any nice way.

It's essential to understand that the solution cannot be found at the end of the chain of causes and effects, but at the root, at the beginning, at where the real problem lies: unsustainable agriculture itself. Solving climate change and global warming, reducing the use of fossil fuels, even reining in population growth could not in themselves lift the dark shadow that hovers at the end of the dead end path.

Like the sudden revelation in a mystery novel that the kindly and mild-mannered butler was the killer, agriculture in its present industrial form has emerged as a villain that threatens the very civilization that it made possible.

By going down the dead end path we've gotten ourselves into a serious situation, but there's no reason to lose all hope. Our challenge will be to create a sustainable future for our race, one that brings humans back into balance with Nature. Our continued existence may literally hang in the balance.

A Review of Climate Change

Although global warming and climate change are not the main subjects at hand, because of their importance and the amount of confusion that revolves around them, a brief review is in order to put them into perspective as they relate to the role of "Air" in the Earth's environment.

Nearly 200 years ago the French scientist Joseph Fourier discovered that carbon dioxide gas in the atmosphere might help to warm the planet.

Around a century ago, Fourier's theory was confirmed by Swedish physicist Svante Arrhenius. Arrhenius published his work in an effort to explain the Ice Ages. His formula relating the amount of CO_2 in the atmosphere to temperature is still used today.

However, being unable to imagine how much fossil fuels would be used as industrialization kicked into high gear, he grossly underestimated the timescale for CO_2 increases, concluding that it would take 3000 years for carbon levels to double. Most climate models now assume that it will take about a century.[146]

The phenomenon came to be known as the "greenhouse effect" because the presence of the gas causes the atmosphere to work somewhat like an ordinary greenhouse. Like panes of glass, the atmosphere is transparent to rays of visible light. Just as the glass holds in heat, molecules of CO_2 absorb infrared rays and prevent them from escaping back into space. In effect, they capture the heat of the Sun and the result is a warmer planet.

Now we need to understand that CO_2 isn't a bad thing. Plants need it to live, and without a certain amount of it in the atmosphere our planet would be a frozen ball of ice. The problem is not the gas itself, but its over-abundance. Like Goldilocks desire for porridge that was neither too hot nor too cold, over millions of years Nature has contrived to hold atmospheric CO_2 at levels that maintain global temperatures in a range ideal for the complex web of life. One might look at it from the other end and say that the environment has evolved to match existing temperature ranges. In reality, both effects are probably true.

It seems that humans were not the first life forms to alter the climate, as atmospheric scientist Stephen H. Schneider concluded in a ground-breaking 1984 book, co-authored with journalist Randi Londer:

"Even without human intervention … it is obvious that

---

[146] Possible feedback effects could further accelerate the rate of increase; for example, if massive melting of Arctic tundra releases quantities of methane, as appears to be happening. Methane has a far stronger greenhouse effect than ordinary CO_2.

climate exercises powerful constraints over the kinds and numbers of living things that can exist on earth. This is graphically illustrated by the wide geographic differences in the distribution of climate types and life forms of today. But what is probably much less well appreciated is the fact that life, as it multiplied and evolved over the aeons, altered the land, air and seas—enough to have changed markedly the very climatic conditions from which earlier life emerged. In a sense, climate and life grew up together. Climate and life have coevolved, to borrow a term determined by population biologists Paul Ehrlich and Peter Raven. Today the balance of mutual influence between climate and life is shifting radically. If we include mankind's social and technological juggernaut as part of the definition of life, then one side effect of our collective footprint on the face of the earth is significant climate modification."[147]

In past epochs the amount of carbon dioxide in the air has fluctuated, and so has the climate. Over the long term, factors have balanced out. Climate change deniers point out that there was far more CO_2 in the air several hundred million years ago, and that life was thriving then, laying down the coal beds and oil deposits that we are mining and pumping today. What they fail to mention (or do not know) is that the younger Sun was not as bright then as it is now. Additional greenhouse gas was required to maintain a balance suitable for life.

At other times, when greenhouse gases have sunk lower, Ice Ages have resulted. It has been clearly demonstrated over more than two billion years of Earth history that the greenhouse effect plays a role in global temperatures and climate.

---

[147] S. H. Schneider and R. Londer, "The Coevolution of Climate & Life," Sierra Club Books, 1984. Schneider was deputy director of the Advanced Study Program at the National Center for Atmospheric Research, Boulder, CO. He died in 2010.

The Earth may be very sensitive to changes in CO_2 levels, as suggested by the co-authors of an unofficial report commissioned by the United Nations and published in 1972. Addressing the subject of the greenhouse effect, they wrote:

"In the field of climate, the sun's radiation, the earth's emissions, the universal influence of the oceans, and the impact of the ice are unquestionably vast and beyond any direct influence on the part of man. But the balance between incoming and outgoing radiation, the interplay of forces which preserves the average global level of temperature appear to be so even, so precise, that only the slightest shift in the energy balance could disrupt the whole system. It takes only the smallest movement at its fulcrum to swing a seesaw out of the horizontal. It may require only a small percentage of change in the balance of energy to modify average temperatures by 2° C. Downward, this is another ice age; upward, a return to an ice-free age. In either case, the effects are global and catastrophic."[148]

Although we still are not one hundred percent certain how much more greenhouse gas will tip the planet's temperature, according to the International Panel on Climate Change (IPCC) the limit of 2° C is the highest it can be allowed to go without disastrous consequences. Unfortunately, in its latest report[149] it estimates that average global temperatures may rise between 1.1° C. and 6.4° C. That's a broad range, but anything much above the low end would be problematic, and the top end would be absolutely disastrous. It's not reassuring to know that the IPCC predictions are generally thought by climate scientists to err on the side of caution.

[148] Barbara Ward and René Dubos, "Only One Earth: The Care and Maintenance of a Small Planet," W.W. Norton & Company, Inc., 1972.

[149] "Fourth Assessment Report," International Panel on Climate Change, 2007.

And if that doesn't get your attention, more recent predictions hint that warming will more likely be in a higher range. Despite loud noise from deniers about the "cooling planet," that just isn't true. Through 2007, the eleven warmest years on record occurred during the previous thirteen years. Unusual volcanic activity and a new phase of El Nino may have led to a pause in the warming during 2009, but if so that's no reason to be complacent, for in fact at the end of 2009 it was ranked the fifth hottest on record. In the United States, January, 2010 was the hottest such month ever, causing record snowfall in some areas.[150] Climate scientists warn that warming may soon return with a vengeance.

In South Australia, which had been suffering through more than a decade of drought, 2009 was the hottest since records were begun in 1910, according to the country's Bureau of Meteorology. A news release[151] reported that the mean temperature for the region was 1.3° C. above average. For the entire first decade of the 21st century the temperature averaged 0.9° C. warmer, "continuing a steady increase in temperatures since the 1970's," the agency reported.

Reports such as that hint that it is entirely possible for average global temperatures to exceed 2° C. in the not distant future, and perhaps by a considerable margin. What would that mean for planet Earth?

Author Mark Lynas addressed that question in a recent book for which he reviewed thousands of scientific studies. He laid out the scenarios at each level of increase up to 6° C. Here is an

---

[150] You read that right: Warmer temperatures allow the air to hold more moisture, and when it meets a cold front, heavy snowfall is the result. To put this in focus, remember the old saying "it's too cold to snow."

[151] "Warmest year on record for South Australia," news release, Australian Bureau of Meteorology, January 5, 2010.

excerpt in which he summarizes the possible effects of an increase of just 4° C.:

> The four degree world will hardly be recognizable. Oceans will continue to rise, perhaps by many meters, engulfing up to a third of Bangladesh and threatening areas around the globe that are home to billions. The rising sea levels will continue for as long as ice remains to melt; in past eons when the Earth was four degrees warmer than it is today, there was no ice anywhere. Today the East Antarctic Ice Sheet alone contains enough frozen water to cause sea levels to rise by more than 50 meters [165 feet]. Of course, that is not going to happen in the near term — but a four-degree world will place us firmly on track for the total disappearance of polar ice.
>
> Everywhere the civilized world we know will be crashing, and nowhere more than in China. With four degrees of warming, China will face feeding its population now approaching 1.5 billion with only two-thirds of the present agricultural production. World markets for food will have disappeared, and in fact are already stretched in the face of the biofuels insanity presently diverting food to fuel. Mass starvation will be occurring as former bread baskets turn into newly minted deserts.[152]

Lynas also describes the disaster that would occur should global warming reach the upper limits of the IPCC estimate at +5° and +6° C., but you probably don't want to hear about it. (If you like horror stories, read his book.) Suffice it to say that Lynas prefaces his discussion of a 6° C. increase by saying that if it were a

---

[152] Mark Lynas, "Six Degrees: Our Future on a Hotter Planet," National Geographic Books, 2008.

TV program it would carry a warning that "viewers may find some of the following scenes upsetting."

In reality, we have only a general idea about what warming of 6° C. would mean, but almost every climate scientist agrees that it would cause a large part of the planet to become uninhabitable.

As a point of reference, Lynas reports that the largest mass extinction in geologic history, the one that ended the Permian era about 250 million years ago, followed a rapid temperature rise in the range of ... wait for it! ... 6° C.[153]

Rising CO_2 Isn't the Only Problem

We hear mostly about carbon dioxide as the greenhouse gas responsible for global warming, but CO_2 isn't the only one. In fact there are many others, some far more potent than CO_2 (but fortunately less abundant). The fourth assessment[154] of the International Panel on Climate Change issued in 2007 listed 18 greenhouse gases, including methane (CH_4), nitrous oxide (N_2O), and CFC-11, better known as the chlorofluorocarbon Freon™, the refrigerant blamed for destroying ozone in the upper atmosphere.

The greenhouse gas that is second only to CO_2 is methane. Methane is the main component in the natural gas we use to heat our houses, dry crops, and produce nitrogen fertilizer. It's also present in enormous quantities in the frozen tundra of the far north, in Canada, Alaska and Siberia. And, those regions are

---

[153] If like most Americans (including me), you think in terms of Fahrenheit, not Celsius, an increase of 6° C. would equal a rise of 10.8° F. As a point of reference, summer growing conditions in the U.S. farm belt often reach the range of 95-105° F. Should those expected highs rise by about 11 degrees, even for a few days, growing corn plants would be seriously damaged at best or more likely completely destroyed. The Farm Belt might become a new Sahara.

[154] IPCC, op. cit.

thawing and releasing the methane back into the air, from which Nature devoted thousands of years to sequestering it in peat bogs.

The Arctic is warming faster than any other place on the planet, and the resulting release of methane could cause global warming to accelerate, as hinted at in this excerpt from an article[155] in *Science* magazine:

> ... widespread permafrost melting could have grave consequences well beyond the far north. No one knows exactly how much carbon is locked up in boreal and alpine permafrost, but estimates range from 350 to 450 gigatons--perhaps a quarter to a third of all soil carbon. The big question is what will happen if even a fraction of this massive carbon store is liberated.
>
> Many parts of the Arctic are already warming faster than any other region on Earth is, a trend that climate models predict will continue. Although researchers are struggling to arrive at a bottom line, they suspect that thawing permafrost will drive global temperatures higher over the next century. No one has a clue, though, how much higher. Thawing permafrost "is a real wild card in the carbon cycle," says Lawson Brigham of the U.S. Arctic Research Commission in Fairbanks.

The tundra isn't the only potential source of methane. It's estimated that thousands of gigatons of the gas are sequestered on the sea bottom in the form of frozen methane hydrate. This ice-like substance is extremely fragile—it melts easily and floats in water. Many of these frozen deposits are located in relatively shallow areas of the oceans, poised to release an enormous amount of methane should the oceans grow warmer—which seems to be

---

[155] Eric Stokstad, "Defrosting the Carbon Freezer of the North," news report, *Science*, 11 June, 2004. A gigaton is a thousand million or one billion tons.

occurring. Should a tipping point be reached in which large amounts of methane hydrates are released, global warming could suddenly accelerate.

One way in which that could take place is through the possible reversal of the Atlantic "conveyor," the currents in which warm tropical water is transported to the region beyond Western Europe and Scandinavia. The warm Gulf Stream carries heat north on the surface of the Atlantic, warming the nearby land. This explains why Ireland and Great Britain, located as far north as Siberia, aren't frozen lands. In response, cold and dense Arctic water sinks deep in the ocean and flows South.

There's evidence that in past climate change events the conveyor may have reversed, and some scientists fear that might happen as a result of global warming. In fact, there are hints that the conveyor is slowing down as further warming takes place in the Arctic. If it should flip over, cold surface water will flow south and warm water will be flowing along the bottom of the ocean, potentially causing a catastrophic melt-down of methane hydrates.

Species Can't Respond to Rapid Change

One issue that gets too little attention is the fact that rapid change leaves environments unable to adapt quickly enough. Deniers claim that plants and animals could easily adapt to higher temperatures, but they fail to take into account that periods of time are required for adaptation to take place.

If a certain species of tree, for example, can no longer survive in its original habitat, but needs to migrate elsewhere perhaps hundreds of miles away, how will it do that in a few decades? It cannot and so will face extinction. And if in order to adapt, a life form must change through evolution, that can take hundreds of thousands or even millions of years.

Global warming is taking place on a scale of decades. It takes a hundred thousand decades to add up to a million years, so you can see how far apart these timescales are. If there's too little time for necessary adaptation or evolution to take place, extinction becomes inevitable—and extinction can occur in a mere blink of time.

It wasn't until quite recently that we became fully aware of the dangers presented by the release of greenhouse gas and other pollutants into the atmosphere. That fact was recently driven home to me when I leafed through a book from my library that I thought might have helpful information for my research. Published in 1970, its title is *Our Precarious Habitat,* and it purports to offer a review of the human impact on the environment.[156]

This is a hefty book, 362 pages with extensive notes and a detailed index. But as I thumbed through this volume, published just 40 years before, I soon discovered a striking fact: Virtually none of the problems we now see as the most serious threats to our environment were even on the radar screens as recently as 1970, at least not as reflected in this book. Here are a few examples:

- The only mention of carbon dioxide in the entire book is in connection with industrial illness from exposure to the gas; not the slightest awareness of its potential effect as a greenhouse gas is mentioned;

- There's no hint of the possibility of climate change as a result of global warming;

- The author does not discuss chlorofluorocarbons and their ability to destroy the ozone layer in the atmosphere;

- No recognition is made of potential sea level rise as a

---

[156] Melvin A. Benarde, "Our Precarious Habitat: An Integrated Approach to Understanding Man's Effect on His Environment," W.W. Norton & Co., 1970.

result of melting glaciers and ice sheets around the world, or of possible changes in ocean currents;

• The terms "rain forest," "greenhouse gas" and "methane" do not appear in the extensive 26-page index;

• Although an emphasis of the book is on public health concerns, nowhere in the index do the terms "Ebola," "Marburg Virus," "Legionnaires' Disease," "Swine Flu," "Lassa Fever," "AIDS," or "Avian Flu" appear;

• The issue of growing water shortages and aquifer depletion is not discussed;

• The author does not touch on the issues of species extinction on land or in the sea;

• Rather startlingly, the words "famine," "starvation" and "malnutrition" do not appear in the index;

• No attention is paid to the subject of resource depletion and the resulting problems for civilization; and finally,

• Population growth is mentioned not as a problem in itself, but only in the context that to feed growing numbers would require increased use of chemical pest control with resulting pollution of the environment.

It was a real eye-opener to see that, in some circles at least, many of today's key issues were not even being discussed as late as 1970. It demonstrates how much awareness of the fragility of our environment has changed in just a few decades.

Much of the book is devoted to discussion of such things as the dangers of industrial pollutants, DDT, radioactive fallout, carcinogenic food additives, bacterial contamination, acid rain, and air pollution. Although those are serious concerns, they are relatively benign effects that can be and in many cases already have been successfully addressed. Environmental issues have moved far past that stage, and at an astounding rate. Today we're well aware of the many far more serious issues noted above that did not appear in the 1970 book on threats to our habitat, not to mention others too many to mention.

Other Effects of the Air

Enough said about how our atmosphere affects global warming and greenhouse gas; now let's look at the other ways in which the "element" I have called "Air" fits into nature.

We have seen in Chapter Five how wind-blown seeds of Emmer wheat were scattered, causing the new species of grass to spread rapidly. That's an example of how the distribution of seeds and pollen are major factors in the reproductive cycle of plants. Bees, wasps, hummingbirds and other creatures of the air also pollinate many flowering plants, serving an essential need in the natural environment.

Bees are of particular concern, because early in the 21st century they began to disappear all over the world due to a mysterious effect dubbed "colony collapse disorder." Efforts to find a cause have been frustrated by the fact that there are apparently many reasons.

According to an article in *The Guardian* newspaper's website,[157] "Disturbing evidence that honeybees are in terminal decline has emerged from the United States where, for the fourth year in a row, more than a third of colonies have failed to survive the winter."

---

[157] "Fears for crops as shock figures from America show scale of bee catastrophe," by Alison Benjamin, Guardian.co.uk, May 2, 2010.

The *Guardian* piece stated that U.S. scientists have identified 121 different pesticides in bee pollen, wax and bees themselves, "lending credence to the notion that pesticides are a key problem."

Bees are an important part of Nature's ecology, and even industrial agriculture cannot do without them. The busy insects are responsible for pollinating plants that are the source of about a third of everything we eat, including most fruits and vegetables, nut trees, sunflowers and oilseeds, coffee, soybeans, clovers, and even cotton. There are 2.4 million hives of bees in the U.S., and in each of the past three years about a third have died. This could represent an important tipping point in the world's ability to produce food.

This spring, a friend of mine who lives in central Missouri reported that he had seen no honeybees in his yard. A few days later he examined a nearby field of blooming clover, walking a grid over about a 100-foot square. He said in an email that he saw a few bumblebees, but not a single honeybee. A blooming field of clover without honeybees is an ominous signal that something very serious is wrong with our environment.

In today's industrial agriculture, the actions of the wind and even pollinating insects can be unwanted effects. When a field is planted to a genetically modified (GM) crop, fields located downwind are sometimes fertilized by pollen from the GM variety. This can result in reduced biodiversity as dominant traits are spread and incorporated into other varieties. Wind can also spread plant disease pathogens, hop-scotching from one field to the next, one nation to another.

It may seem reasonable to use standard, non-diversified crops, but in fact it creates a serious potential problem. When a new plant disease, fungus, or other threat appears, it affects different varieties within a plant species to varying degrees. Some plants may be extremely vulnerable, while others resist the attack. Diversity is Nature's insurance policy against the future.

What would happen if our world, supporting a bloated and still growing population on the slender legs of a few undiversified crops, should be faced with a pandemic of plant disease? Cereal grain crops—primarily wheat, rice and corn (maize)—provide about half of the calories consumed by human beings worldwide. That means that the world food supply could be seriously threatened if a new plant pandemic should appear that targets one of these three major food crops.

Because of intensive plant breeding that has produced the high-yielding varieties of the Green Revolution, most diversified crops with their differing qualities of disease resistance have been replaced by one-size-fits-all varieties. That leaves a dispropor-tionate amount of the worldwide grain crop potentially susceptible to a pandemic of disease that could spread from nation to nation with disastrous effects.

Is it reasonable to think that such an event could happen? Well, if you're familiar with history you may know of the Irish Potato Famine of the 1840s, and that will give you some idea of the possibilities. Ireland had depended on its potato crop as a main source of food, much as the entire world today relies on wheat, rice and corn. In Peru, where the potato was first domesticated, there are more than 50 varieties of potato. In Ireland pretty much a single variety was grown, the ubiquitous Irish spud, white, plump, and starchy.

In the late summer of 1845, a virulent leaf blight almost completely destroyed Ireland's life-sustaining crop. Spread by the wind, the blight had a devastating effect on the Emerald Isle, with a population of about eight million. The crisis continued for several years, resulting in an estimated half-million deaths, displacement of two million Irish refugees, and the emigration of another two million to other countries such as the U.S. and Canada.

But that was just in one small country, and affecting a crop that was none too important in most other parts of the world. What would happen if something similar should overtake one of the world's major food grains? In fact, you may not have read about it in your daily paper or seen it on network news, but in 1999 something like that appeared—a plant fungus of a type called "stem rust," and one that was deadly to wheat.

Wheat rust wasn't something new. Different strains of the fungal infection had caused significant grain losses and famine early in the 20th century. As recently as 1950, an outbreak of stem rust destroyed nearly 70 percent of the wheat crop in North America. One of Norman Borlaug's important achievements in launching the Green Revolution was to develop wheat varieties that were resistant to then-prevalent strains of stem rust. Now, like a zombie rising from the grave, this ancient enemy of *Homo agriculturis* was back on the march.

After ten years of frantic effort, by 2009 a resistant variety had been identified and 1.5 tons of seeds produced, only sufficient to produce more seed. But we weren't out of danger yet, for the disease still has the potential to become a full-blown disaster of worldwide proportions. This is an unfolding story.

Known as Ug99, the new plant disease was first identified in Uganda in 1999 (thus the name) and began to rapidly spread. In 2005 plant scientists from 18 countries met in Kenya to address the problem, an event called the Global Rust Summit. A summary report from the meeting stated that because Ug99 "has broken down the source of stem rust resistance that has protected much of the world's wheat for 30 years, the crop is poised for an epidemic to spread like wildfire."

By 2015 UG99 had been joined by no less than 10 different strains of wheat rust, and the plant plague had spread along the entire east coast of Africa, from South Africa to Egypt, as well as to the Arabian Peninsula and as far as Afghanistan.

According to a report on the website of the journal *Nature*:

> "Eventually it will reach North America and Europe," says Ronnie Coffman, a plant-breeding scientist at Cornell University in Ithaca, New York. He warns that in the next few years, farmers across the world will need to replace up to 90% of the current wheat varieties with new, resistant varieties to ensure crops are protected against the fungus.[158]

One of Norman Borlaug's crowning achievements in Mexico during the 1950s was to introduce genes into his hybrid wheat varieties that made them able to resist rust. With the emergence of Ug99, that protective shield has been pierced. In all, seven variants of Ug99 have been found, according to the *Nature* report, and several of these are more dangerous than the original Ugandan strain. This is ominous news.

A screening at the USDA research station at Beltsville, MD, of about 2000 wheat varieties used in the United States revealed that more than 70 percent were vulnerable to Ug99, including 82 percent of spring wheats, 69 percent of hard winter wheats, and 73 percent of soft winter wheats. Data from Canada showed that the majority of about 100 Canadian wheat varieties were also susceptible. And even worse, it turns out that many varieties of barley and oats are also subject to infection with Ug99.

---

[158] "Virulent wheat fungus invades South Africa," Nature.com, May 26, 2010.

As can be seen, even if resistant strains can be found and planted, wheat diversity will be severely limited as a result of the many varieties that are vulnerable to Ug99. Besides loss of diversity, how many other valuable traits will disappear—such things as higher yield potential, ability to use water more efficiently, resistance to other diseases and pests, and the ability to stand up to the next plant pandemic? It is possible, if not likely, that many of the advances of Green Revolution plant breeding may have to be thrown away in favor of the ability to resist Ug99 stem rust. Meanwhile, the virus continues to mutate and change to defeat the ability of wheat varieties to combat the infection.

By June, 2009 the problem of Ug99 had become serious enough to begin to attract public attention. An article[159] in a leading Canadian newspaper reported:

> Scientists in Canada and around the world are racing to find a way to stop a destructive fungus that threatens to wipe out 80 per cent of the world's wheat crop, causing widespread famine and pushing the cost of such staples as bread and pasta through the roof.
> Canadian officials say that the airborne fungus, known as Ug99, has so far proved unstoppable, making its way out of eastern Africa and into the Middle East and Central Asia. It is now threatening areas that account for more than one-third of the world's wheat production and scientists in North America say it's only a matter of time before the pest hits the breadbasket regions of North America, Russia and China.

The newspaper quoted research scientist Rob Graf, who said: "I think it's important people start recognizing what a big threat this is. This could mean world famine. This is quite the deal."

---

[159] "Fungus Threat Hangs Over World Wheat Production," The Ottawa (Canada) Citizen, June 25, 2009.

Quite the deal indeed. In a world that has become almost totally dependent upon cereal grains, the idea of a plant disease pandemic is beyond unsettling. Already we have seen the steady spread of avian flu, which although it may never morph into a human pandemic is already undermining an important source of protein in human diets by endangering chickens, ducks, turkeys and other domestic fowl. For example, in the spring of 2015 avian flu was spreading across the U.S., affecting laying hens and turkeys. This was pushing up the cost of meat and eggs, and there were reports of egg rationing in some areas.

There are a host of other plant and animal diseases rising up, and the more we become dependent upon highly selected mono-cultural crop varieties and standard animal strains, widespread famine has become a real possibility. In fact, unless some significant changes take place, it's not a question of whether, but when.

What Could World Famine Be Like?

How serious could such an event be? Using the Irish Potato Famine as a model, what if a disaster of similar proportions should strike the entire world today, virtually destroying the wheat crop everywhere? Yes, that's a worst-case-scenario, but not entirely unrealistic because it's based on what actually happened in Ireland. It's interesting to project the possible scale of the disaster if something similar might occur on a worldwide scale.

In the 1840s, half a million of eight million Irish people died. In today's world, where about 2 billion people depend on wheat, that would represent about 125 million deaths. Besides that, two million Irish became refugees, and another two million emigrated — about half the original population in all. Here's where the problem becomes really serious, because on a global scale involving about two billion people dependent upon wheat for a large part of their diet, that would translate to fully a billion people.

But those figures for displaced persons and immigrants cannot be extrapolated because there would be no place for a billion people to go in the face of worldwide famine. The Irish were able to go elsewhere to escape the famine. In case of a global pandemic of wheat rust, there would be nowhere to go. In our present example, we must conclude that those billion would also starve.

And, we're not done yet, because our worst-case model calls for all wheat to be destroyed by the pandemic, and thus the entire cohort of 2 billion wheat-dependent people could be considered at risk of death, nearly the combined populations of China, the U.S., Indonesia and Brazil, the world's first and third, fourth and fifth most populous nations.

The preceding comparison is no doubt far-fetched, but it does provide some indication of just how serious a major disruption of the global food supply could be. If the scenario were only one-tenth as bad, there could still be 200 million victims.

With all these new threats blowing on the wind, the world could soon face some real difficulties, possibly causing the era of our time to be viewed in future histories as a time of great famine.

Chapter Eight

The Third Element: Fire

The "element" I call "Fire" is energy, and without it our planet would be a sterile, frozen ball of rock. In a state of nature, virtually all energy comes more or less directly from the Sun. The Earth itself does not produce energy—even the heat of lava, exploding volcanoes, earthquakes, and tsunamis merely result from the release of energy already in the Earth. The forms of energy used in the cycle of Nature virtually all derive from the Sun.[160] Solar rays drive the winds and climate, causing rain to fall and rivers to flow. Plants use photosynthesis to create growth. Herbivores consume the plants and derive energy from them. Carnivores consume the herbivores. Bacteria, nature's recyclers, busily work to process dead plant and animal matter for future growth. All depend on the power of the Sun.

There are two ways in which energy is used in agriculture. The first is created from food in the muscles of humans and animals such as plow horses. This is called metabolic or

---

[160] It can be noted that ocean tides play a very minor role. They result from the gravitational attraction of the Moon.

"endosomatic"[161] energy. This is the kind of energy you use to walk, run, and lift objects, and it has its source through our food by the process of photosynthesis by plants. In earlier times, virtually all the energy used in farming came from muscle power.

The Sun is a reliable source of energy; it is the epitome of "renewable" but has its limits. Solar energy flows at only one speed—you can't "turn up" the light from the Sun to produce more from an acre of land. It can work only over time and through the cooperation of Nature, through sustainable cycles of growth.

Because solar rays must be captured by plants, their use as a source of muscle energy cannot easily be scaled up. For example, in traditional agriculture if a farmer wants to double the number of acres he could plow using one horse, a second horse must be added. To feed that horse requires more acres for pasture and to grow fodder and grain. A team of two horses would only incrementally increase the number of acres that could be plowed because not all time is spent actually moving the plow and two horses would be harder to handle than one. The farmer would have to increase the time spent in the fields, caring for the animals, growing and harvesting grain and hay, and maintaining their harnesses and tools. He might need to add a hired helper. Such are the limits of muscle power.

There are other forms of energy that the Sun has made, laid down in the Earth over hundreds of millions of years. Those are the fossil fuels—petroleum, coal, and natural gas, all derived from ancient life forms, products of photosynthesis.[162]

---

[161] Literally, "coming from inside the body."

[162] A few scientists speculate that oil and natural gas may have come from mineral sources deep in the Earth. This idea is not widely accepted. Coal is clearly derived from fossilized plant matter.

Using Energy from the Earth

These fuels provide what is called "exosomatic" energy, that coming from outside the body, and in their most common forms they are gasoline, diesel fuel, natural gas and coal. The use of exosomatic energy can be scaled up quite easily—no need for an extra horse or new pastures. Machines burn fossil fuels to perform work[163] that human or animal muscles would have performed in the past. Machines can do far more work, and do it faster, thus explaining their popularity.

Here's a simple example: a small gasoline engine can convert 20 percent of the energy in a gallon of gasoline into power. That yields 8.8 kilowatt hours of energy.[164] A human can do work equal to about one-tenth of one horsepower. A 12-hp gasoline engine running for one hour will produce mechanical work that would take a human 120 hours to perform.

If a little engine can do that, think how much more a giant diesel tractor can perform. For example, in 2014 John Deere Co. introduced its model 9620R farm tractor, featuring a 14.9-liter diesel engine rated at 620 h.p. It carries 400 gallons of fuel and can travel at up to 25 mph. Based on horsepower alone, that single tractor is capable in one hour of doing work that would take a human nearly three years of 40-hour weeks (6000 hours). Unlike the 620 live horses that would be needed to replace it, the machine

---

[163] I am using the word "work" as defined by the science of Newtonian physics, i.e., mechanical work, measured by "the amount of energy transferred by a force acting at a distance." Except for the motion of my fingers, what I am doing at my keyboard is not "work" by this definition. What's happening inside my head doesn't count. However, because clever ideas can reduce the amount of mechanical work required to perform a task, humans can expand their productivity beyond the merely physical. Machines cannot, at least, not yet.

[164] M. Giampietro and D. Pimentel, "The Tightening Conflict: Population, Energy Use, and the Ecology of Agriculture," posted at www.dieoff.org, 1994.

doesn't require additional acres to grow feed; its exosomatic energy comes from the ground in places like Saudi Arabia, Venezuela and Iraq.

That's an incredible magnification of the amount of work that can be done by a single farmer sitting in an air-conditioned cab and steering a green-and-yellow mechanical monster. By such examples, it's easy to see why the use of fossil fuel has allowed industrial agriculture to expand as rapidly as it has.

By 1994, in developed countries more than 40 times more energy from fossil fuels was being used than from traditional muscle power. In the U.S., the world's greatest user of energy, the ratio was 90 to 1.[165] In rough terms, compared to pre-industrial agriculture, it means one farmer can do 90 times more work using industrial methods.

Should the supply of oil run out we might conclude from this that the number of farm workers would have to be increased by 90 times to equal the present level of production. There are around 6 million farm workers in America today, two percent of a total population of about 300 million. A 90-fold increase in farm labor would equal 540 million—nearly double the entire population including bankers, bakers, butchers, bellhops, burger-flippers—every one of us, even babies in their cribs.

It would require 180 percent of our population just to grow food for the 300 million we actually have, an obvious impossibility.

This is admittedly a simplistic observation—machines can do things that humans cannot, and any return to an agrarian model would require huge changes from present methods—but it does give us some indication of the scale of the problem. It's interesting

---

[165] Ibid.

to compare that back-of-the-envelope calculation with the fact that prior to the beginning of the Industrial Revolution "only" about 95 percent of Americans lived and worked on farms.

These facts could lead us to conclude that agricultural production was actually more efficient 250 years ago when humans, not machines performed most of the work. That might even be true in a way since people bring brain-power to the task, but farming was done on a much smaller scale and with a much larger per capita base of fertile land to be exploited. However, it seems logical to conclude that while capable of vast amounts of work, machine farming may not be particularly efficient.

Can We Return to Nature?

If without oil it would be necessary for the majority of Americans to be employed in agriculture, the truth is that there may not be enough land to sustain such a return to an agrarian society by our present population. That's because high crop yields are dependent upon heavy use of fertilizer and ag chemicals, which are derived from fossil energy.

For example, we've seen that corn yields in Indiana are more than eight times higher than in 1930 before industrial agriculture began to take root. Imagine what it will mean when fertilizer and ag chemicals are no longer affordable or even available. If the world can barely feed its billions today, should crop production suddenly fall by seven-eights it would be a disaster almost beyond our imagination.

Other solutions must be sought, but it's not going to be easy because we have traveled so far down the dead end path that our unnatural methods seem "normal" to us now. As the Green Revolution demonstrated, we tend to seek solutions through more industrialization rather than through new approaches to food. This flies in the face of the fact that resource depletion will eventually cause industrialization to stop.

As I build the case against technological agriculture, one of the most important bodies of evidence revolves around the substitution of fossil fuels for natural processes. We have seen that by using fossil fuels to replace human labor, population has been able to expand far beyond any sustainable level. That could not have happened except by the systematic depletion of the Earth's stores of coal, oil and gas.

Now we face a double-edged sword of consequences. First, the release of CO_2 from fossil sources is warming the planet, creating a clear and present danger to civilization. Second, the supplies of those unnatural energy sources are peaking with possibly dire results.

Without the intrusion of fossil-fueled technology, Nature would have remained in balance. There would be no more plants and animals—including humans—than the natural processes of the planet could support. That balance held for eons until humans learned to farm by digging up the soil, and most importantly, since we discovered how to vastly magnify our efforts using energy drilled or dug from the Earth.

We have seen that even in its primitive form agriculture has been a growing problem for humans, as attested by the collapse of ancient civilizations. But it is only since we started using fossil fuels, and in particular during the boom that began after World War Two, that the possibility has emerged for potentially disastrous famine.

Original, non-industrial agriculture would have reached its limits as the population spread over the available land. At the point where no more new land could be settled, natural constraints would have come into play and balance would have been restored.

That was the conclusion of Malthus, who pointed out that population in his time was growing faster in the New World than in England where the limits of Nature were already established. Of course, as we also saw, he failed to foresee that industrialization would allow humanity to jump far beyond the natural bounds of what the Earth can support.

We've seen the differences between exponential population growth and arithmetic increases in food production. Until about 250 years ago, population expanded in accord with Malthus's predictions. It was when exosomatic energy from fossil fuels began to replace muscle power that we made a fateful commitment to the dead end path.

Statistics show three related factors that seem to run in tandem, with population growth, use of fossil fuels, and CO_2 emissions all three beginning to soar upward after the advent of the Industrial Revolution until today all three are climbing almost straight up on the vertical axis of any graph.

As we have seen, population had been on a steady, expo–nential increase until the time of Malthus around 1800. The growth was able to continue by the opening of the New World and the general expansion of human numbers made possible by traditional agriculture.

But at about the time he expected that rate of increase would begin to slow due to the balancing effect of Nature, human numbers began an even sharper rise. The parallel tracks reveal the reason, because they track carbon emissions and the amounts of fossil fuel being burned. That line remained more or less level until a little over two hundred years ago at the start of the Industrial Revolution. Then, the lines for both population and emissions take off like the trajectory of a Saturn rocket heading for the Moon.

The similarity to a Hubbert bell curve is obvious. In fact, any

non-renewable or unsustainable resource can be plotted on a
Hubbert graph. Besides oil and other natural resources such a
graph may also reflect Peak Food, Peak Population, and Peak
Civilization. The use of unnatural energy has taken us down, or
should we say "up," the dead end path. But where do we go from
here? A vast expanse of future post-resources time looms ahead of
us, and to survive as a species we will eventually have to deal with
it, or it will certainly deal with us.

Today's Agriculture Depends Upon Fossil Fuels

Energy is a keystone element in the operation of industrial
agriculture. In 2002 it was estimated that the following forms of
exosomatic energy were used by U.S. farms:[166]

Electricity 20.7%
Gasoline 8.5%
Diesel Fuel 27.3%
LP Gas 4.5%
Natural Gas 3.6%
Fertilizers 29.0%
Pesticides 6.3%

Direct use includes such things as gasoline and diesel fuel used
to operate cars and pickups; farm machinery used to prepare,
plant, and harvest crops; and trucks to transport supplies, crops
and livestock to and from markets. Propane and natural gas are
used for heating and to dry wet-harvested crops. Electricity is used
for lighting, heating, operating milking equipment and other
systems. None of these were available to pre-industrial farmers,
whose exosomatic energy use was limited to burning wood or coal

---

[166] Randy Schnepf, "Energy Use in Agriculture: Background and Issues,"
Congressional Research Service, The Library of Congress, order code RL32677.

to cook, heat their dwellings, do smithy work, or smoke meat. For the better-off farm family, perhaps there were a few candles or lanterns to shine against the night.

This list also includes two categories of indirect energy, used to make fertilizers and pesticides. As can be seen, those two categories alone accounted for slightly more than a third of farm energy use.

In all, this direct and indirect use on farms amounts to only about 1.7 percent of all energy used in the U.S.—but that doesn't take into account all the many other ways in which the food chain uses energy. In 1994, food-related activity accounted for about 17 percent of total energy use in the U.S., roughly in line with the percentage of average income spent on food.

We have seen that in a very real sense oil and food are joined at the hip. As we pass beyond Peak Oil and fuels become increasingly scarce and expensive, so with food. As one writer puts it, we are "eating fossil fuels." In his book[167] by that name, journalist Dale Allen Pfeiffer makes essentially the same case I am presenting here, that we are using up irreplaceable resources to produce food. He concludes: "Sustainable agriculture for all intents and purposes means a return to small-scale farming, where the acreage can be managed by a family and a horse or mule with a plow."

Would that it were so simple. As we have seen, there is not nearly enough land to feed the present world population through traditional farming, much less a surplus of horses and mules. A population that has grown so far beyond long-term sustainability absolutely requires fossil fuels for civilization to exist in its present mode in the United States and other "developed nations. Thanks to Green Revolution style farming we have gone far beyond the stage where a return to a simple agrarian past is a reasonable

---

[167] Dale Allen Pfeiffer, "Eating Fossil Fuels," New Society Publishers, 2006.

option, at least certainly not in any short term and not with the numbers of people alive today.[168]

In the "developing nations" of the world, perhaps to their good fortune, the oil-machine model has not yet become dominant. Most such places are still in the late stages of traditional agriculture, and have yet to rush down the dead end path.

In China, for example, a fair percentage of the population remains on small farms. The Middle Kingdom has an enormous population—more than 1.393 billion as of June, 2015. At that time the country had a labor force of 853.7 million, of which abut 43 percent were employed in agriculture. That calculates to about 367 million farm workers—more than the total population of the United States. Clearly, endosomatic energy is still a significant factor in Chinese agriculture.

If China were to industrialize its food production to the extent the United States has, reducing their farm labor force to around 2 percent of the population, it would more than 300 million peasants unemployed and seeking jobs elsewhere. Needless to say such a thing would be impossible, so China is left no option but to continue to support a less industrialized form of agriculture.[169]

Developing Areas Won't Get Off Easy

Even though developing countries may not have as much invested in mechanized agriculture, that doesn't mean they are immune to the effects of declining energy resources and rising prices. In fact, the poorest people of the world are the most likely to be harmed.

---

[168] It's worth a reminder that "development" is a code word for exploitation, so what does that say about our "developed" nations?

[169] Nonetheless, China seeks to industrialize the rest of its economy, creating a kind of schizo, class-divided parody of a modern "civilization" having more in common with earlier slave-based economies. But that's another issue entirely.

Here's what I wrote on my blog back in 2008 when oil prices were spiking upward and consumption was falling in response:

> But since world oil consumption is actually declining along with supply, what else might be at play here? Well, one thing for certain is that poor people everywhere, whether in America or in the Third World, are being affected the most. They are the most vulnerable, the first to feel the pain. Poor and failing nations are already priced right out of the world's oil market; they just plain cannot afford it. For that reason, the price of water buffaloes has risen by 80 percent in the Philippines just in the last few months, and in some Arab regions the supply of camel meat (a popular foodstuff) has fallen dramatically (and prices have risen) because people are keeping the camels for transportation instead of sending them to slaughter.
>
> And it is not only oil for transport and other direct energy uses, but also indirect knock-on factors that are hitting the economies of the Third World a deadly blow. The startling rise in food costs is in large part a reflection of higher oil prices (although the collision between population and the ability of the Earth to produce food is a major factor). With both energy and food prices soaring, the situation is shaping up as a major disaster in the Third World and we should expect serious consequences there as substantial numbers of people pass over the line between malnutrition and outright starvation.[170]

---

[170] "Why Less Oil Demand Won't Reverse High Prices," *Star Phoenix Base*, www.starphoenixbase.com, June 3, 2008. I was addressing the long-term view. While prices did decline substantially in the short term, it was due to a worldwide recession rather than sudden increases in oil supply. In the longer term, diminishing supplies of oil will be bid up and only the richest will be able to afford it at all. When I wrote the first edition of this book in mid-2010, oil was ratcheting up and down in the $70-90 level. At the end of 2015 it was hovering around the $40 level due to excess supply and declining demand. I suspect we'll

Higher oil prices also translate into higher costs for fertilizer and farm chemicals, and for subsistence level farmers who can barely put food on their own table, that's a catastrophe. During 2009 there were hints that some farmers everywhere—even in the United States—were using less of those inputs because of the high costs involved. The inevitable result would be lower yields. There were reports of some who were planting nothing at all, being unable to afford the expensive hybrid seeds, fertilizers, and farm chemicals being forced on them by globalized agri-businesses, another "benefit" of the Green Revolution.

It is interesting to see early hints of a trend back to animal rather than machine power. The farmers in the Philippines and Arab teamsters in the Middle East had not traveled far down the dead end path, so it's still possible for them to step back. Farming by water buffalo, or transporting goods by camel instead of by truck are long-established methods in those areas.

But imagine if you can a typical American. farmer with a thousand or so acres of corn to plant, grow, and harvest—and unable to afford diesel fuel for his machinery. Could he consider a return to farming with a team of horses as my Grandfather did in the 1940's? No, of course not. For one thing, there are few draft horses left, and for another most farmers wouldn't have any idea how to manage them. At best, a team of horses would suffice for 40 acres or so, not a thousand. What may work in the Philippines and Saudi Arabia just wouldn't fly in Indiana, Iowa and Illinois.

As the Oil Peak sets in, we hear much about alternative fuels. These include wind, solar, geothermal, wave and biomass

continue to experience a yo-yo effect, with rising oil prices causing a series of recessions to cause demand (and prices) to drop once again. Remembering that oil = food, spreading famine is a likely consequence as this plays out.

sources. Unfortunately, most alternatives are ill-suited to replacing gasoline or diesel fuel. Most of them are used to generate electricity, something that is almost impossible to store, difficult to transport, and unable to provide energy off-the-grid except through the use of storage batteries. Considering those limitations, try to imagine battery-powered tractors or combine harvesters capable of turning out two or three hundred horsepower, hour after hour during a critical planting or harvesting season. It would take a lot more energy than batteries could provide—technology that we not only don't have, but which will likely never exist.

Even electric automobiles such as those coming to market now raise troubling questions. For one thing, the cars themselves are a product of industrial technology, so just to manufacture them uses up more resources. Furthermore, they add to the demand for electricity.

One estimate showed that if even a relatively small percentage of Californians switched to plug-in cars, the capacity of the state's power grid would be exceeded during peak charging hours at night, causing brownouts or blackouts. That would call for construction of more power capacity, another technological use of resources. As we've seen in other examples, this is moving in the wrong direction, toward more use of technology, not less.

Since more than half of electric plants burn coal, some wag has suggested that plug-in electric vehicles are actually coal-burning cars. There's more than a little truth in that observation.

One of the problems with the present oil supply situation is that increasingly less desirable sources are being tapped, things such as tar sands, oil shale, and heavy, high-sulfur oils. These require a lot of processing, perhaps consuming almost as much energy as they yield, or even more. The only reason these resources are being developed is to replace "real" petroleum such as light crude that is no longer available in sufficient quantities. What happens is that other forms of energy are expended to produce liquid fuels that can be stored, transported, and used on demand— substitutes for gasoline and diesel fuel.

The False Alternative of Bio-Fuels

The same limiting factors apply to bio-fuels as they are presently being produced by more than 200 ethanol distilleries scattered across the nation. Not only do ethanol producers bid up the price of corn, they also remove the grain from export markets, thus indirectly punishing poor nations that are unable to produce enough food for their people. Many foreign countries rely on American corn exports, and in 2008 when the prices of grain rose in tandem with petroleum, it forced up food prices around the world. In 2015 corn prices settled back to around four dollars a bushel, again in tandem with dropping oil prices. Ag economists estimated that to be close to the breakeven price for Midwest farmers. Good news for ethanol producers; bad news for farmers.

According to a World Bank report issued in the summer of 2008: "Biofuels have forced global food prices up by 75%—far more than previously estimated."[171] Not only does the practice disrupt grain markets, it isn't even very efficient. Some studies have shown that it may take more energy to produce a gallon of ethanol than it contains, and as the price of corn is bid up the cost of ethanol rises with it. Ethanol distillation also uses a lot of electricity, most of which comes from coal or gas fired plants.

In fact, ethanol producers face a Catch 22 situation. When oil

---

[171] "Secret report: biofuels caused food crisis," *The Guardian* on-line edition, July 3, 2008.

prices are low, corn prices also tend to decline. That's good for the distillers because they can buy the corn more cheaply to produce alcohol. But it's also bad, because when gasoline is cheap the pump price for ethanol will also be low. When oil prices rise, corn prices will trend up, increasing ethanol production costs.

Despite that, in 2014 more than 14 billion gallons of fuel alcohol were produced in the United States, according to the industry's trade organization, the Renewable Fuels Association.[172] Nearly all was made using corn. A similar process was being used to make fake diesel fuel from soybeans, something called "bio-diesel." When Congress allowed a subsidy on bio-diesel to expire at the end of 2009, many bio-diesel plants shut down, clear evidence that the practice of making fake fuels from farm crops is economically unsound unless supported by government largesse. Biodiesel production has continued to spike and plunge in tune with the expiration or reinstatement of federal credits.

At the farm production end of this questionable practice of making corn into alcohol, farmers will be encouraged to plant ever more crops, requiring more fuel, more fertilizer, more chemicals, and leaving more soil subject to erosion. The production of ethanol is the runaway Frankenstein Monster of mechanized agriculture at its worst, swallowing its own tail in a vain effort to maintain the unsustainable model of industrial farming, running faster and faster down the dead end path in an effort eventually doomed to fail.

We need to understand that ethanol is not gasoline, but a

---

[172] The very name of the trade association hints at the misleading arguments made in support of ethanol, because it is in no way "renewable." It is in fact only another way through which to profit from the depletion of resources, including soil and energy. I will stop short of calling the ethanol boom a scam or Ponzi scheme, but there are obvious similarities. Let's just say it's a misguided, profit-oriented endeavor.

pretender that disguises itself as the "real thing". There are plenty of ways in which it falls short. For one, alcohol attracts and absorbs water. When mixed with gasoline, as it usually is, it causes the blend to evaporate more quickly, making long-term storage a problem. It can dissolve plastic, varnish and rust, leading to contaminated fuel.

It also takes a lot of water to produce ethanol, about three gallons for each gallon produced—and that's just at the distillery. If it's made from corn grown on irrigated fields, who knows how much water is used? As we'll see in the next chapter, water is a major factor in the twisted picture of today's industrial agriculture.

Then there's the fact that ethanol contains less energy than gasoline, a lot less. If you think you can substitute ethanol one-for-one for gasoline, you'll have to think again. When Consumer Reports compared the performance of a 2007 Chevrolet Tahoe Flex-fuel SUV using E85, a blend of 85 percent alcohol, versus straight gasoline, "Overall fuel economy on the Tahoe dropped from an already low 14 mpg to 10."[173] The test report also mentioned that the vehicle's range dropped from 440 to 300 miles, meaning more frequent fill-ups. Since ethanol is priced as equal to gasoline, but cannot do as much work, it is obviously less economical. The only advantage to all this seems to belong to the investors in ethanol plants. Of course the corn growers may think it's to their benefit, but they are merely the tools of corporate entities, mining their fields and accepting all the cost and risk of the scheme.

Imagine if you will a farm itself running on ethanol fuel while growing corn ultimately used to produce more ethanol. Talk about a dog chasing its own tail. First the farmer must burn fuel in his

[173] *Consumer Reports*, October, 2006.

machines to plant the field. He must buy hybrid seeds grown elsewhere, harvested, transported, processed, bagged and delivered to the farm,[174] all requiring more energy inputs for the crop. Then he must pay for nitrogen fertilizer made from natural gas feedstock, and apply pesticide, herbicide and fungicide made from petroleum. Next the crop must be tended through the growing season, utilizing more energy and perhaps repeated treatments with ag chemicals. Then, it's time for harvest and the big combine burning even more fuel.

But it doesn't end yet. Trucks burn even more fuel to haul the moist grain out of the field, perhaps to an on-farm dryer where natural gas or propane is used to dry it for transport. The grain is then taken to an elevator and put on rail cars for transport to the ethanol plant. There it's processed (using energy), heated (more energy), fermented and distilled (using, well you get it), then sent by tankers or pipeline to a distribution center where it's blended with gasoline to produce E85. From there it goes back on the road in tank trucks for delivery to the farm, where Spring is approaching and it's time to plant another crop. The unnatural cycle of destruction begins anew.

Is there something wrong with this picture? Yes, there is. Ethanol is an unnatural fuel because it eats up the very resources it's intended to replace. When oil became the King of Energy, it

---

[174] It may surprise you, but in 2009 farmers paid as much as $250 or more for a 50 lb. bag of patented, genetically superior hybrid seed corn. (Source: *Agronomy Advice* newsletter, by Joe Lauer, U. of Wis. agronomist, August, 2009) That's enough to plant up to about four acres, so the investment required just for the seed to plant a thousand acres could amount to as much as $62,500 if top varieties are planted The cost is justified by increased yield. If corn brings $3 a bushel, an additional yield of 21 bushels per acre would cover the expense of the seed and any more would benefit the farmer. Note that the seed corn companies are in line first, as is always the case in our present model of industrial agriculture where farmers are always left holding the short end of the stick.

was plentiful and flowed out of the ground almost by itself. There was no repeated need for external energy inputs to produce the end products.

And finally, if ethanol is such a good idea, then why does the United States government have to subsidize the stuff? It's true—ethanol producers receive generous kickbacks from Uncle Sam for every gallon they produce, while valuable farm crops are turned into a poor substitute for oil. Some three-quarters of all government payments in support of alternative fuels goes to ethanol subsidies, an amount that was expected to reach $5 billion in 2010. That's more than the USDA would spend on conservation programs to protect soil, water and wildlife habitat. The federal subsidy on ethanol in 2009 was 51 cents per gallon—but there is an additional "hidden" subsidy because the government allows ethanol to be priced the same as gasoline, even though it contains considerably less energy. The total subsidy has been estimated at well over a dollar a gallon.

Two agricultural scientists from different parts of the world summed up the problems with ethanol in a letter to the editor[175] of *Science* magazine:

> Although biofuel's contribution can be positive, it will remain small, being restricted by the ability of the natural environment to provide both fuel and food for a large and energy-demanding world population.
> It requires production equivalent to 0.5 ton of grain to feed one person for one year, a value sufficiently large to

---

[175] David Connor, University of Melbourne, Australia, and Inés Minguez, Universidad Politécnica de Madrid, Spain, *Science*, 23 June, 2006. Note that a car attaining 33.5 mpg on ethanol would perhaps be able to travel about 50 mpg on gasoline with its greater energy content.

allow some production to be used as seed for the next crop, some to be fed to animals, and some land to be diverted to fruit and vegetable crops. Compare this value with that for a car running 20,000 km/year [12,500 miles] at an efficient consumption of 7 liters/100 km [33.5 mpg] The required 1400 liters of ethanol [370 gals.] would be produced from 3.5 ton grain (2.48 kg grain/liter), requiring an agricultural production seven times the dietary requirement for one person.

In the same issue, three scientists from the University of Vermont noted: "…the entire state of Iowa, if planted in corn, would yield approximately five days of gasoline alternative." They added: "We need alternative energy. But ethanol from corn is neither scalable nor sustainable. Let's pursue better options."[176]

There are claims that ethanol can be made from things other than corn, substances such as crop waste, wood pulp, or purpose-grown switchgrass. All of these have problems, and in the end you're left with corrosive, low-energy alcohol — and the planet is left the worse. We've seen the importance of crop residue to build up organic matter in soil. And estimates of how much energy it takes to produce ethanol from switchgrass ranges as high as 150 percent of the amount of energy the end product contains. None of these are useful even as short-term solutions.

'Nuf said about ethanol. But what about nuclear power? Isn't that the hope of the future? Well, back in Chapter Three when we discussed resource peaks, I didn't mention uranium because I wanted to keep that as a surprise. No, we're not about to run out of the stuff tomorrow—but there are definite limits to how much of it

---

[176] Nathan Hagens et al., *Science*, letters to the editor, 23 June, 2006.

we can expect. According to a *Wall Street Journal* story[177] that appeared in late 2009, it may not be too soon to start worrying about Peak Uranium. Here's a take-away from the article:

> The fears over "peak uranium" boil down to simple math: The world presently consumes a lot more uranium than it produces. The latest numbers from the International Atomic Energy Agency [IAEA] say global annual consumption is 69,100 tons; global production from mining is around 43,000 tons. The difference—for now—is basically made up from nuclear-weapons stockpiles, which obviously aren't an infinite resource.
>
> That's the arithmetic that has renewed "peak uranium" chatter in recent weeks. Swiss scientist Michael Dittmar talks of a supply crunch as soon as 2013. And all those worries are based on the size of the world's current nuclear power fleet.
>
> The thing is, China, India, the Middle East, and the UK are already ramping up their own nuclear renaissance. The U.S.—the world's largest user of nuclear energy—has plans for more... Either way, nuclear expansion on the drawing board seems likely to increase the world's appetite for uranium.

The report also says that the IAEA insists there is an 80-year supply remaining, but remember the economics of the Hubbert curve—that once the halfway point is passed, that is, the resource peak is reached, supplies will begin to fall and prices go up. It is unlikely that any resource will be completely extracted because it will become too expensive to do so. Demand will eventually fall to zero because the resource will be priced right out of sight. In that regard, the back side of the Hubbert curve is a model of the law of diminishing returns as rising demand chases diminishing supply.

---

[177] "Peak Uranium: More Reasons to Worry About Powering the Nuclear Revival," *Wall Street Journal* on-line edition, December 10, 2009.

An 80-year supply implies total depletion, and as always a Hubbert curve suggests Peak Uranium will no doubt come much sooner. In the United States, once the world's largest source, uranium peaked in 1980, according to the Organization for Economic Cooperation and Development (OECD). And, as we have seen, nuclear power is useful primarily as a source of electricity, which would hardly be a substitute for diesel fuel or gasoline.

Hydrogen Is Not the Answer

Another potential alternative is hydrogen, in its molecular form H2. In fact, we hear glowing predictions of a future "hydrogen economy," in which our cars and other machines will burn pure hydrogen, emitting only water. The unfortunate truth is that this is another pipe dream aimed at maintaining the status quo by finding a replacement that will let us continue to operate as we always have, rather than seeking truly different ways. It's a plan to continue down the dead end path by other means.

Hydrogen cannot be substituted for petroleum products. It can be liquefied only at high pressure and at temperatures below minus 252.87 degrees C (-423.17 degrees F). It is highly flammable (as witness the fate of the dirigible Hindenburg) and cannot be easily stored or transported. Our present distribution systems for liquid fuels and natural gas are unsuitable for a hydrogen economy, and thus an entirely new infrastructure would have to be built at great expense. Many of the machines of our industrial technology would become obsolete and need to be replaced as well.

Now there's another Catch 22 relating to hydrogen—in fact, we might call it Catch 22-squared. It's a chicken-or-egg thing, for what comes first, hydrogen powered cars, trucks, tractors and other machinery, or the vast infrastructure to supply hydrogen fuel? Without the full development of the one, there is no rational justification for the other. And development of one is impossible unless the other already exists. It's unreasonable to think that our oil-based technologies in all their complexity could be easily and seamlessly replaced by an entirely different infrastructure.

And, oh yes, it takes a lot of energy to produce hydrogen by breaking the oxygen bond of water. Where would that energy come from? Some suggest solar energy, but the cost of a nationwide solar-hydrogen system would be immense. And on a related note, the huge infrastructure for an H2 economy would have to be built using conventional energy such as oil and gas, placing even more demand on dwindling supplies. Except for limited, *in situ* applications, hydrogen is a non-starter as a replacement for our present fossil fuels which are transportable, efficient, and for which we have a complex infrastructure representing trillions of sunk dollars.

What other energy options are there for the post-Peak Oil era? Well, they're really kind of thin on the ground. There are hopes for a new kind of reactor using thorium instead of uranium. Thorium (Th) is a slightly radioactive element with an atomic number of 90 (uranium's number is 92). A mix of 232Th and 233U has been successfully tested in nuclear reactors. It's estimated there is about four times more thorium in the ground than uranium.

But still, that returns us to the problem of energy delivered as electricity when the real need is for a concentrated and easily stored and transportable energy source—something like gasoline or diesel fuel. Plus, like so many other "alternatives," this is another example of using up ("developing") a non-renewable resource.

Fusion Power: A Will-o'the-Wisp

Finally, there is the ever-hopeful promise of fusion power, which has been touted as the ultimate solution to all of our energy problems—if only we could figure out how to make it work. For more than 50 years we've been told that fusion power is "just 30 years away," and it remains there today, ever receding into the future like a will-o'-the-wisp. One thing is certain, and that is that fusion power will not be available in a time frame that will do much good for our present crisis as the Oil Peak unfolds. Even 30 years is probably entirely too late.

If fusion power ever does come into being, it will no doubt be expensive and, like other sources of electricity, it wouldn't be a practical source of energy to drive tractors and other farm equipment. And, as ever, it's more than likely that fusion power plants would require huge amounts of external resources to build and operate—resources that are ever-dwindling.

The bottom line is that industrial agriculture depends on concentrated, high-energy liquid fuels. It is dependent upon oil. Are there any real alternatives on the horizon? Well, let me leave you with one potentially bright spot, and that is the possibility of producing fuel from algae. In 2009 ExxonMobil committed to invest $600 million in a joint-venture algae fuel program with Synthetic Genomics, owned by Craig Venter, the first person to map the human genome.178 Unfortunately, after four years and spending $100 million, Exxon backed out of the deal saying that it would take 25 years to develop the technology on a large scale.

Still, green algae has great promise as a source of real alternative fuel. First, it will not be used to produce ethanol. The algae has a high content of natural oils that can be refined to produce products virtually identical to diesel fuel and even gasoline and jet fuel.

---

[178] "Oil Giant Exxon Sees the Future—and It Is Green Algae," *The Independent*, on-line edition, July 15, 2009.

The algae can be grown in tanks without endangering and "mining" farmland. In fact, the ideal place for an algae farm is in the middle of a desert where there's plenty of sunshine. The stuff grows rapidly in recyclable water, and bless its little green heart, it grows by gobbling up loads of CO_2 (which is unfortunately released when the fuel is burned, but at least it's a zero-sum game).

There are several promising algae projects under way. One, by Sapphire Energy of San Diego, CA, is pioneering what it calls "green crude oil," created from renewable algae. In late 2009 the U.S. Departments of Energy and Agriculture gave the company grants and loan guarantees worth more than $100 million to develop the process. In the pilot plant being constructed in New Mexico, the "green crude" will be used to make biodiesel and jet fuel. The company had already demonstrated the concept and sponsored several airliner test flights powered by fuel made by the company. Going into 2010, experiments were underway to prove the ability to use "green crude" to make gasoline.

Still, there are problems to be overcome because to be efficient algae farms must make minimum use of fertilizer and fresh water. One suggestion is to build algae facilities next to wastewater treatment plants or power plants that produce CO_2.

Algae may provide a link that could begin to replace oil and natural gas with truly renewable fuels in the longer term, ones that use the natural processes of photosynthesis rather than drilling and digging up irreplaceable sources that will soon be in short supply, or depleting fragile and precious soil. In the end, even algae biomass energy is unsustainable because it uses up nonrenewable resources.

And, according to a recent article in *New Scientist* magazine,[179]

---

[179] "Biodiesel from algae may not be as green as it seems," *New Scientist*, 24 July, 2010.

the technology of producing biodiesel from algae needs to be more efficient to be ecologically sound.

Energy, the "Fire" component of the cycle of nature, is a crucial element in agriculture. There are few substitutes for the powerful fossil fuels that have taken us for a dangerous ride down the dead end path. There may be other alternatives, but the honest truth is that nothing we have yet imagined could equal West Texas Light Sweet Crude, flowing effortlessly out of the ground.

Industrial agriculture is completely reliant on fossil fuels, and as those resources reach their limits, we cannot hope for very much longer to continue our present style of food production. However, by developing renewable sources such as from algae, we might provide stepping-stones toward a different future, to give the human race time to adapt to the need for sustainability. It's not too early to begin planning for a post-Oil and post-Gas future. In fact, it's far too late, but as the Roman historian Titus Livius[180] wrote more than 2000 years ago: "Better late than never."

---

[180] Livius, known as Livy in English, is best known for his classic "History of Rome".

David L. Brown

Chapter Nine

The Fourth Element: Water

Water, an oxide of hydrogen (H_2O), is one of the most common molecules in the universe. This isn't surprising since hydrogen and oxygen are the first and third most abundant chemical elements.

Water is ubiquitous on and near the surface of our planet and is the foundation for life. Seventy percent of the Earth is covered with oceans and the cycles of rain and snow bring this life-giving substance to all parts of the land except the most barren deserts. Without water, our planet would be as barren as the Moon.

Water is the fourth, and arguably the most important "element" in the cycle of Nature that brings forth the plant life that provides sustenance for the animal kingdom.

Water is a renewable resource, recycled through the process known as the hydrologic cycle. This is the process through which water is evaporated from the ocean, carried over land by the wind, falls as rain or snow, and is captured in mountain glaciers or flows down streams and rivers before returning to the ocean to complete the cycle. The energy driving this constant renewal comes from the Sun.

But despite its presence almost everywhere, in many places fresh water is becoming a vanishing resource in the face of industrial agriculture and the growing numbers of human beings. About two billion people now live in areas with insufficient water, and unless changes are made, by 2025 two-thirds of the world's population will suffer from water shortages.

If water is a renewable resource and in large supply, how can water shortages be a problem? One answer lies in the fact that not all water is equal. For purposes of life, fresh water is required. Remember the complaint of the Ancient Mariner adrift in the salty sea: "Water, water, everywhere, nor any drop to drink."[181]

Unfortunately, most of the Earth's water is saturated with salt and other compounds that make it impossible to drink and toxic to plants. All but about three percent of the Earth's water is found in the oceans, and two-thirds of the rest is bound up in glaciers and ice sheets. Only about one percent of water is available for human use.

Even that might seem bountiful, but fresh water is not found everywhere. Fully 20 percent of it is contained in Lake Baikal in Asia. Another 20 percent is found in the Great Lakes of North America. Flowing rivers account for only .006 percent of the world's total water and much of the rest is found in underground reserves, many too deep to be reached.[182] Rainfall varies widely, from flood-inducing tropical monsoons to the tiniest hint of occasional rain in places like the Sahara and Gobi Deserts.

Human water use has expanded dramatically in lock-step with

[181] Samuel Taylor Coleridge, "The Rime of the Ancient Mariner," a poem, 1798.

[182] "The Water Cycle: Freshwater Storage," U.S. Geological Survey web site http://ga.water.usgs.gov/edu/index.html, updated as of June 10, 2010.

industrial agriculture. During the 20th Century global population tripled, but water use increased by 700 percent. The Green Revolution depended on copious use of flood irrigation to produce the bumper crops that "saved" billions from famine. The amount of land under irrigation tripled after 1950.

In an ideal situation sufficient water falls from the sky as rain. That's true in some fortunate places such as the American Corn Belt, but in many other regions there's not enough natural rainfall to sustain modern agriculture. One need only think of the desert areas of California where irrigation is used to grow many of our nation's vegetables and salad greens (and which is beginning to reach the limits of water availability).

Irrigation is one of the primary tools of industrial agriculture, bringing millions of acres of formerly unproductive land under cultivation. This isn't something new—the very first farmers must have learned to water their gardens and early agrarian civilizations in the Fertile Crescent and along the Nile Valley built vast networks of canals and channels for irrigation. In our era of industrial farming water use has been vastly multiplied through the use of fossil fuels to drill, pump and deliver the precious liquid.

There are three basic kinds of water resource that can be tapped for irrigation: flowing rivers; lakes and impoundments; and underground aquifers. All three are seriously threatened by over-use, pollution, and declining quantities.

Rivers Are Being Sucked Dry

All around the world, rivers are being pumped dry for the production of food. Thus the Colorado River often fails to reach the Gulf of California in Mexico, because all of its waters have been used up. Climate change and over-use are accelerating this problem along major rivers around the world.

Not only the Colorado, but also the Rio Grande in America, the Ganges and Yangtze in Asia, the Nile in Africa, the Jordan in the Mideast, the Murray-Darling in Australia, and the Po in Italy sometimes fail to reach the sea. Some are in danger of disappearing altogether.

Rivers are fed by rainfall, springs and wetlands, and in the case of many of the largest, from melting snow and glaciers in high mountains.

According to Fred Pearce, a British journalist who covers the environment, rivers provide the largest proportion of the world's fresh water. Unfortunately, those rivers are being overtaxed. Investigating the failing river resources around the world,[183] Pearce reports:

> About 32 billion acre-feet of water makes the journey from the land to the sea every year. Or it did before we started diverting it. So how much of this runoff do we use. Of those 32 billion acre-feet a year, much hurtles off the land in occasional floods or flows away from permanent rivers. Of the rest, hydrologists estimate the maximum that might reasonably be caught and used by humans employing current technology is 11 billion acre-feet. But nature has played one more trick on us. Many of the world's greatest rivers are in regions where few people can or want to live. The three rivers with the biggest flows—the Amazon, the Congo, and the Orinoco—all pass through inhospitable jungle for most of their journey from headwaters to the sea. These three alone carry almost a quarter of the water we have to survive on. And two more of the top ten—the Lena and the Yenisei, in Siberia—run mostly through Arctic

---

[183] Fred Pearce, "When the Rivers Run Dry: Water—

the Defining Crisis of the Twenty-First Century," Beacon Press, 2006.

wastes. A tenth of the world's river waters flow into the Arctic. Take out these and we are left with around 7 billion acre-feet of river water for our needs

That still amounts to about 370,000 gallons a year for every citizen on the planet. Not bad, but I calculated my own annual water use at between 400,000 and 530,000 gallons a year. I imagine most of the world would like to live as well as I do. So we have a problem.

As mentioned, many rivers have their headwaters in high mountains where glaciers and winter snows melt to provide a steady flow of water. This applies to the Colorado and Rio Grande in North America, most of the major rivers of Asia that come down from the Himalayas, and the Rhine, Danube, Po, and Rhone in Europe. Unfortunately, due to climate change many if not most of the world's glaciers are receding.

As they do, the rivers they feed begin to dry up. I remember my own experience upon revisiting a Swiss valley where in the early 1980's I had photographed a massive glacier hanging above the cliffs. Returning in 1997, I was surprised to see that the glacier had receded out of sight into the high mountains.

Not only are rivers being starved of water at their sources, but human intervention has affected the waterways by straightening their courses, draining wetlands, building levees, and putting up dams. The general result is to turn the riparian environments into little more than a series of pools and open channels running straight to the sea.

And that's not the worst, because rivers have been used for generations as convenient depositories for human waste, toxic chemicals, and other pollutants. The Ganges, sacred river of India, is little more than an open sewer, filled not only with waste but also with human corpses consigned to the waters. Although many of the dead are cremated before being placed in the river, it's not uncommon to see bodies floating in the stream. Nearby, children may be splashing in the holy waters, others bathing and brushing their teeth.

It's not only in the developing world that rivers have been made toxic. The Mississippi that flows through the heart of America carries runoff from farm fields that contains fertilizer, agro-chemicals, and other pollutants. When the water flows into the Gulf of Mexico, each summer a so-called Dead Zone is created, caused by algae overgrowth. The algae grows out of control thanks to nitrogen and phosphorus in the runoff, depleting the oxygen in the water. This is a direct side-effect of industrial agriculture. In recent years the Dead Zone has reached an extent of up to 8000 square miles, about the size of New Jersey.

Lakes Are Sinking, Too

The second source of water for irrigation and other use comes from lakes and impoundments behind dams. Like rivers, these waters are being over-used, or diminishing due to climate change or upstream withdrawals. The story of the Aral Sea in Uzbekistan is a classic example. Once the fourth largest freshwater lake in the world, it has nearly dried up thanks to a Soviet-era scheme to use the waters of the river that feeds it to grow cotton. This inland sea, which once covered 26,000 square miles and was the source of a rich fishing industry, is today a virtual wasteland.

Again, it's not just in faraway places that surface water is disappearing. In the dry Southwestern United States, huge reservoirs created on the Colorado River are also dropping. The second largest man-made reservoir in the United States is Lake Powell, created by Glen Canyon Dam. The lake can hold 30 cubic kilometers of water.

Unfortunately, due to water use and drought, the level has

been dropping and the ability of the Colorado to refill it is seriously compromised due to drought. The flow of the rivers feeding Lake Powell have run below average for some years now and in June, 2015 the reservoir held just 48 percent of its capacity.

Further downstream the even larger reservoir, Lake Mead behind Hoover Dam, is also shrinking, reaching near minimum levels in recent years. In April, 2015 it reached the lowest level since May, 1937 when it was first being filled and stood at just 37 percent of its capacity. As these impoundments recede it raises serious concerns for the region, which relies on them not only for water, but also for the electricity produced by the Hoover and Glen Canyon hydroelectric dams.

The scale of the growing problem is reflected in this excerpt from a news report:[184]

> There is a 50 percent chance that Lake Mead, which was created by the Hoover Dam and the Colorado River, will go dry by 2021 because of escalating human demand and climate change, according to a study by Tim Barnett and David Pierce of the Scripps Institution of Oceanography of the University of California at San Diego.

The report continued:

> By 2017, there is a 50 percent chance that the reservoir could drop so low that Hoover Dam could no longer produce hydroelectric power. Water conservation and mitigation technologies and policies thus need to be implemented now, the study stated.
>
> The disappearance of the manmade lake would create

---

[184] Michael Kanellos, "Lake Mead May Go Dry by 2021," CNET News, news.cnet.com, February 17, 2008. The scientific report was set to appear in *Water Resources Research*, a journal of the American Geophysical Union.

a tidal wave of ill effects for the southwestern U.S. The lake provides water for large cities like Los Angeles and Las Vegas, as well as for several agricultural interests. The power also keeps on the lights in that region of the country. Imagine Los Angeles on a summer day with sporadic air conditioning and only a trickle of water coming out of the faucet.

There are many other examples of lakes that are receding or even disappearing, everywhere from Asia and South America to Africa and Australia. Here are excerpts from reports that provide specific examples:

- "**Lake Victoria** in Africa [Uganda], the world's second largest freshwater lake, has been experiencing receding water levels since 2001. As it recedes, more than 30 million people who depend on it for livelihood are facing a disaster."[185]

- "Evaporation attributed to global warming is drying up **Lake Titicaca**, one of the world's highest navigable lakes. Its water, which straddles Bolivia and Peru 12,493 feet above sea level, has reached its lowest level since 1949, authorities said last week. They blame the water loss on the rainy season, which has been reduced from six to three months."[186]

- "The waters of the **Sea of Galilee** where Jesus is said to have once sailed are at their lowest level on record due to drought and demand from Israel, it has been claimed. The freshwater lake which supplies Israel with much of its drinking water and irrigation has been

[185] J. L. Awange et al., "Observing Our Changing Earth," Springer Berlin Heidelberg, 2008.

[186] "Lake Titicaca Drying Up Because of Warming," Climate Wire, posted November 16, 2009 on the Earth News website, www.earthportal.org/news.

hit by a combination of four years of drought and relent-
less demand from homeowners and farmers in the
region."[187]

- "Long considered a national treasure in Mexico,
Lake Chapala in the state of Jalisco is shrinking so fast
that many people here worry about damage to the
environment and the Mexican economy. Since the
1970s, scientists say, the lake has lost about 80% of its
water due to heavy development in central Mexico. The
lake is fed primarily by the Rio Lerma, which meanders
through several hundred miles of arid farmland and
supports about 11 million people along its banks.
Farmers in recent years have taken to diverting almost
all of the river's flow to irrigation, often with outmoded
techniques that experts say use up much more water
than necessary. At the same time, the bustling manufacturing center of Guadalajara lies downstream and
draws on the lake as its principal source of water."[188]

These excerpts were included in the first edition (2010) and
there have been some reversals. For example, Lake Chapala has
regained much of its water during the recent El Nino event.
However, Lake Mead in 2015 reached the lowest level since the
1930s when it was first filling with water.

'Mining' Water from Beneath the Ground

When natural rainfall is insufficient to sustain crops and
there are no handy rivers or lakes from which to drain water,
industrial farmers probe beneath the Earth to tap groundwater. In

---

[187] "Sea of Galilee Water Level at Lowest on Record," Telegraph.co.uk, August
29, 2008.

[188] Jim Carlton, "Shrinking Lake in Mexico Threatens Future of Region," *Wall
Street Journal*, September 3, 2003.

many cases these are "fossil" deposits that are refreshed slowly over long periods of time. Thus, the deposits of water are being used up. Although they are theoretically renewable, the rate of replacement in many cases is measured in hundreds or even thousands of years.

Essentially, when pumped out these resources are gone forever as far as humankind is concerned. The parallel of aquifer depletion with the "development" of petroleum is striking.

An excellent example is provided by the Ogalalla Aquifer, which is a vast underground resource in central North America. The aquifer extends over portions of eight states in the western High Plains of the United States.

Also called the High Plains Aquifer, it covers 174,000 square miles and provides 30 percent of American groundwater used for irrigation. The formations that contain the water were formed two to four million years ago during the late Miocene and early Pliocene geologic eras. Some water is replaced through natural rainfall, but the process is slow and most of the remaining water has been in the ground since the last Ice Age. It's estimated it could take 2000 years or more for the water to be restored from rainfall soaking into the deep earth.

As we saw in Chapter Six, when settlers first attempted to farm the High Plains the results were the Dust Bowl and financial disaster. The failure came from trying to use conventional farming methods in a region lacking sufficient rainfall, under the mistaken belief that "rain will follow the plow." There was water there, beneath the dusty land, but beyond the reach of the technology of the early 20th Century. When the native prairie sod was broken, there was nothing to hold the soil down before the drought and winds.

The use of industrial methods to create deep wells, pumps, and sprinkler irrigation systems has created an entirely different story for the High Plains. Today, thanks to machine technology those once-failed lands are under intensive cultivation. Ironically, it is the power of fossil energy that was put to work to draw fossil water from deep underground and sprinkle it onto growing crops. Without technology and fossil fuels, corn, alfalfa, grain sorghum and other thirsty crops could not be grown in many parts of the dryland region.

The most common method of industrial farming in the High Plains is based on center pivot irrigation. In this system a well-head is located at the center of a field from which a long pipe on wheels with sprinkler heads spaced at regular intervals rotates like the hour hand of a clock, sweeping out a circle of irrigated soil. Fly over this region and you will view one of the strangest sights on earth, with thousands of crop circles dramatically illustrating the intensity of technological agriculture in the pell-mell destruction of non-renewable resources.

As you can readily realize from the fact that the aquifer does not contain an infinite amount of water, and that only a small proportion of what is pumped out of the ground is replaced, this form of industrial agriculture is unsustainable. In just a few decades, wells have sucked the equivalent of the annual flow of 18 Colorado Rivers from the aquifer.[189] Some estimates indicate that at least parts of it could be depleted in as little as 15 years, and that is especially true in the southern parts of southwest Kansas, New Mexico, Texas and Oklahoma where the least water remains.

Meanwhile, wells must be dug deeper and more energy used to bring the receding water to the surface. The aquifer is not in the form of underground lakes, but is contained in formations of sandstone or shale. Thus, when a farmer pumps out two acre-feet

---

[189] Maude Barlow, "Blue Covenant: The Global Water Crisis and the Coming Battle for the Right to Water," The New Press, 2007.

of water, and if there is a one-to-ten ratio of water to stone, the water table itself might drop by as much as twenty feet. This is approximately the amount of water needed to grow a single crop of high-yielding corn.

A Worldwide Phenomenon

Unfortunately the story of the Ogalalla Aquifer is far from unique. Supplies of fresh groundwater are rapidly declining everywhere in the world. As with every other natural resource on which modern agriculture depends, the human race has been drawing down fresh water exponentially.

Underground resources are being pumped from the Earth at a furious rate, in places such as India, China, Vietnam, Mexico and Pakistan. Here are some factoids:

> • In India farmers draw water 24/7 through more than 20 million wells, using deep drilling methods pioneered by the petroleum industry. Another million wells are being added each year. They pump 50 cubic miles of water from the ground each year. In many areas the wells are running dry, and in recent years thousands of Indian farmers have committed suicide as their crops withered and died for lack of water.[190]

> • Northern China, the nation's wheat breadbasket, is turning into a desert as groundwater supplies vanish. More than 7 cubic miles of water are sucked from the land each year, causing the water table beneath Beijing to fall by nearly 200 feet just in the past two decades. The region is plagued by sand and dust storms reminiscent of the 1930s Dust Bowl in the United States.

---

[190] Ibid.

The region is steadily succumbing to desertification.[191]

• In the Punjab region of Pakistan, source of 90 percent of the nation's food, water tables have been plunging. This mountain nation is home to more than 175 million people in an area less than twice the size of California. Less than a quarter of the land is arable. According to the CIA Factbook, Pakistan faces the following environmental issues: "water pollution from raw sewage, industrial wastes, and agricultural runoff; limited natural fresh water resources; most of the population does not have access to potable water; deforestation; soil erosion; desertification."[192]

As human numbers grow and agriculture and other industrial applications use more, More, MORE of the precious liquid, in many areas we are approaching Peak Water. Statistics that sown accelerating fresh water use do much to illuminate the headlong rush down the dead end path that began with the end of World War Two, driven by industrialization. A sudden upward turn occurred about 1950. Thanks to the Green Revolution, the largest amount of water use is attributed to agriculture. Imagine this as the left-hand side of a typical Hubbert bell curve, predicting a peak sometime in the future followed by a steady decline. As with so many other resources, that peak may be coming soon.

The problem with water is that there is often either too much of it (flooding) or too little (drought). It seldom seems to be at the right place at the right time. These problems are becoming more serious as climate change alters seasonal weather patterns in many areas. So while we cannot expect a global Peak Water event, we are already passing such thresholds on local and regional levels due

[191] Ibid.

[192] The World Factbook, "Pakistan," Central Intelligence Agency, www.cia.gov.

215

to the fact that water is not available everywhere it might be used. Many regions that rely on rivers, lakes, and underground reserves are in danger of running out of fresh water. For them, Peak Water is very much a reality.

Quite simply, potentially productive soil and water must be found together in the right combination, and this is seldom the case without human intervention. It is notable that about 40 percent of the world's food is produced through irrigation.

There's another reason why this process is not sustainable, and that is that the continued application of water can literally poison the soil. This occurs when water contains mineral salts that are leached out of the ground before the water is applied to cropland. As the water evaporates, the salts are left behind, building up alkali crystals in the soil.

'Peak Water' Occurs Regionally

On a global basis the comparison of water with Peak Oil is imperfect for several reasons. First, when petroleum is "developed" it's used up and gone forever. Fresh water suffers no such fate; it is a renewable, or rather, recyclable resource, even if on a long time scale in many cases. Oil is valuable enough that it makes economic sense to transport it around the globe, especially when the energy to do so remains relatively cheap. Water is not worth the cost of shipping it, and that is unlikely to change as energy costs rise.

Nevertheless, some areas of the world are "exporting" water to drier places. No, not the real thing, but "virtual water" in the concentrated form of food. For example, grain grown in North America is shipped to places such as Saudi Arabia, Egypt, and other nations that lack enough water to grow their own crops, in effect selling them water that falls from U.S. or Canadian skies or is pumped from rivers or out of the ground. For example, much of the farming that takes place in the High Plains goes for export, with the declining resource of the Ogallala Aquifer being shipped overseas in the form of grain, meat, or cotton. Thus does "progress" continue through the destruction of resources.

The export of virtual water is feasible because growing grain uses a lot of water. For example, a high-yielding corn crop uses about 600,000 gallons of water per acre from planting to harvest. That typically amounts to around 3,000 gallons per bushel of harvested yield.[193] Imagine the transportation savings enjoyed by shipping the corn itself rather than the water needed to grow it. A bushel of shelled corn weighs 56 lbs., and 3000 gallons of water weighs more than twelve short tons.

Amazingly, even water deficit regions engage in the export of virtual water. Ever think about those fresh fruits and veggies you find in the supermarket in February? Chances are many of them came by air from arid places like Peru, Chile, or South Africa where the southern Summer is in full swing. In the U.S., many of our strawberries year around come from Mexico, a nation that is suffering desertification and steady loss of water reserves.

The fact is that much of what we eat comes with many miles on its "odometer." The significance of these "food miles" is that it's another way industrial agriculture uses up energy resources, especially petroleum, by moving food all over the globe. Here's an excerpt from a paper written for Earth Policy Institute:

> Food today travels farther than ever, with fruits and vegetables in western industrial countries often logging 2,500–4,000 kilometers [1500-2500 miles] from farm to store. Increasingly open world markets combined with low fuel prices allow the import of fresh produce year-

---

[193] "It takes a lot of water to grow a corn crop," *Southeast Farm Press*, December 28, 2007.

round, regardless of season or location. But as food travels farther, energy use soars. Trucking accounts for the majority of food transport, though it is nearly 10 times more energy-intensive than moving goods by rail or barge. Refrigerated jumbo jets—60 times more energy-intensive than sea transport—constitute a small but growing sector of food transport, helping supply northern hemisphere markets with fresh produce from places like Chile, South Africa, and New Zealand.[194]

Even transport of food to its final destination, the last stage in technological agricultural, relies on vast amounts of fossil fuel, making it possible to trade "virtual water" in the form of food to every corner of the globe. Never mind that the precious water, like the energy used to transport it, is being rapidly used up in many areas. As petroleum becomes scarcer and more expensive, it will eventually no longer make sense to move food long distances.

What About Desalinization of Sea Water?

One last note on the subject of water concerns the use of desalinization plants to turn seawater into fresh water. This is beginning to see wide use in dry places such as the Middle East. However, desalinization is an expensive proposition and especially if geared toward farm production, because of the large amounts of water required. Most desalinized water today is used for drinking, cooking and bathing.

According to the U.S. Geological Service, desalinized water costs up to five times more to produce and deliver to consumers than fresh water from natural sources such as rivers and aquifers, perhaps $1000 per acre-foot. Of course, those places that have no

---

[194] Danielle Murray, "Oil and Food: A Rising Security Challenge," Earth Policy Institute, May 9, 2005. The author was a staff researcher at EPI.

alternative such as Dubai, Saudi Arabia, and Israel must pay the price. The use of this process is just another example of a technological "fix" that is dependent upon the very resources that are being destroyed.

We've seen how the four "elements" of farm production—fertile soil, favorable climate, energy, and water—are key limiting factors for industrial agriculture. These essential ingredients are being used up, polluted, and despoiled in the race to produce food for a world population that is increasing exponentially, and, of course, in the name of the Almighty Dollar.

As these essentials are used up and destroyed, runaway technological "progress" is carrying us forward toward the terminus of the dead end path. In the next chapters, we'll see how economics and politics have helped to drive this headlong rush to an uncertain future.

David L. Brown

Chapter Ten

The False Prophet's Tale

We've explored the ways in which industrialization, and especially its impact on agriculture, has set the world on a downward path. But the question remains: How in blue blazes have the bursts of "development" and "progress" since World War Two managed to take place? It doesn't seem to make sense, because constant growth and the accumulation of excess "stuff" and "wealth" for their own sakes is irrational.

Economists measure success by GDP growth, as if the only thing that matters is obtaining more, More, MORE of everything. Materialism has become the defining feature of our time. It's like being stuck on a merry-go-round that just won't stop. People want bigger houses, newer cars, wall-sized TV sets, the latest cell phones, more food, finer wines, and enough clothing to fill a haberdashery. It's crazy.

This state of rampant "development" and runaway consumerism has occurred through the manipulation of economies and people by international corporations with the aid of governments, usually for the direct benefit of the "haves" and to the detriment of the "have-nots." It proceeds through the destruction of natural resources and relies upon the illusion that wealth can be created essentially from thin air. It operates with the convenient support of economic theories and policies that are often based more on wishful thinking and fairy dust than the facts of Nature.

All of this frantic activity is certainly not done for the benefit of the consuming public—the creation of all that "stuff" and "wealth" is merely a mechanism through which the true aim is achieved: to further enrich the rich and empower the powerful. The masses are exploited through advertising and other means that addict them to materialism. When there's insufficient wealth to keep the process going, it's accelerated through the use of easy credit, debt, and the printing of money.

In historic times, before the invention of economists, the cornerstone of human affairs was tangible capital. For a pre-agricultural tribe, capital included such things as a comfortable rock shelter to live in, a handy source of flint for tool-making, a place to gather nutritious food, and access to plentiful game. It also included the human capital of the gatherer, the medicine man, the arrow-maker, and the hunter. In other words, capital was something that was found in the natural world and which had real value, both material and human.

Something has changed, and it can be seen in the widening gap between the definitions of "capital" and "wealth." Today's wealth is nothing like that of the past, which derived exclusively from real assets such as land, livestock, control of serfs or slaves, and precious metals. Today economists and politicians have twisted the facts of the universe, creating a false tale of wealth.

Much of what today is called "wealth" is a fiction and fantasy built out of nothing. To understand this, observe what happens when the Federal Reserve turns on its printing presses. Does an influx of "new money" into the banking system increase the wealth of our nation? No, for these pieces of paper do not represent anything real. The increased money supply merely dilutes any capital upon which the "money" supposedly is based.

Something like that happened in ancient Rome, where the value of gold coins was undermined by the addition of base metals. Much the same process of capital dilution is taking place today. In the case of U.S. currency, it's backed by nothing more than the "good faith" of the government. Thus, it's that very "good faith" that is diluted and devalued when more money is issued. We'll take a closer look at the phenomenon of "faith-based currency" in a later chapter.

Note that there is a difference between "real' wealth that is based on true capital, and paper wealth. The owner of land, metals, or other non-depreciating assets controls true capital. The day trader who buys and sells electronic packets somewhere in cyberspace deals in a kind of imaginary wealth, not too different from Monopoly money or the plastic chips in a casino. Most of the "real" capital wealth is owned by the powerful few who control our corporations and governments; the less fortunate of us must be content largely with the imaginary kind.

So what is wealth if it can exist as a fantasy? According to Frederick Soddy, writing nearly a century ago in a little-remembered but insightful book, money is actually debt, owed by the issuer to the holders of the money.195 When money is not backed by a hard asset such as gold or silver, it creates what Soddy

---

[195] F. Soddy, "Wealth, Virtual Wealth and Debt: The Solution of the Economic Paradox," George Allen and Unwin, Ltd., London, 1926; new printing 1983 by CPA Book Publisher, Oregon. Soddy was not an economist, but a Nobel Prize chemist who applied the methods of science to an examination of economics. Although he was considered a crank at the time, many of his economic proposals are now common practice.

called "virtual wealth." This creates difficulty because uncollateralized debt carries a negative, not positive value. Nonetheless, these kinds of debt are often considered as assets and thus added to the supposed wealth of a nation. In other words, debt can be parlayed into wealth, over and over again.

When an economy is based on real capital, it can grow only within strict bounds because there are limited amounts of hard assets to back it. This is inconvenient for those who wish to participate in uncontrolled "progress," the never-ending multiplication of wealth. That's why the world went off the gold standard—its price was held to a minimal value, but there still wasn't enough gold to satisfy the desire for virtually unlimited growth.

Some students of economic history say that the Great Depression was a result of the restraint upon growth by the gold standard during the 1920s. For example, Berkeley professor Barry Eichengreen wrote:[196] "The gold standard limited the control of central banks over monetary policy. As a result, central banks could not lower interest rates in order to expand demand and stimulate the economy, deepening and prolonging the Great Depression." The U.S went off the gold standard in the early 1930s.

When credit enters the picture and money becomes a faith-based commodity, the limits imposed by available capital no longer apply. A central bank can turn to the printing press[197] to create new "wealth," but the amount of real capital remains fixed. Credit thus becomes equated with wealth, even though it is a negative quantity. Bankers and financiers earn obscene bonuses based on

[196] B. Eichengreen, "Golden Fetters: The Gold Standard and the Great Depression," 1919-1939. Oxford University Press, USA, 1992.
[197] In fact, even printing presses are no longer required. The Federal Reserve can simply create money in the form of computer bits and bytes.

the creation of debt through the issuance of loans and other instruments. Central bankers oblige by printing more money as required to feed this insatiable monstrosity.

Resources Are Assigned Little Value

Not only is fictitious "money" evoked to represent wealth, but the future value of the natural resources that are the real capital of the planet are dismissed as having little worth. Hard to believe? Yes, but true. Since economists rig their theories to enable constant "growth," they discount the future value even of non-renewable resources.[198]

This is an absolute requirement for their purposes because to strip the Earth for short-term gain could never be justified if the increasing future value of those resources were to be fully recognized and accounted for along with the harm to the environment that results from their destruction.

To look at this in another way, economists and financiers evaluate investments using the principle of Net Present Value (NPV). NPV is derived from the sum of the values of individual cash flows, whether positive or negative, discounted to present value. Man-made assets such as buildings, factories, ships, and airplanes are typically depreciated into the future. Inconveniently, if left alone natural resources don't generally depreciate and often actually gain in value.

---

[198] When a mineral claim is filed on public lands in the United States, the prospector is allowed to buy the land for as little as $2.50 per acre, and often is allowed to "develop" the resource without paying anything back to the taxpayers who originally owned it. In addition, the government may choose to pay the cost of building a road and other infrastructure for the new mine. And that isn't all, for through "depletion allowances" the owners of natural resources can enjoy tax deductions against the amount of those resources that are developed. All this is done in the name of "progress."

To recognize the dangers of fictitious accounting that inflates present value and profit, one need only examine the history of Enron Corporation and the resulting destruction of the accounting giant Arthur Andersen. The problem has long been recognized. For example, a leading U.S. accountant expressed the following opinion in 1975:[199]

> Present financial accounting does not reflect economic reality. Worse than that, it creates an illusion of enormous profits where often no true profit exists. Our economic system is floundering from a severe lack of real profits essential to capital formation and preservation.[200]

It would be hard to justify much if not most industrial development should natural resources be analyzed with full accounting for future appreciation; the cost of replacement (if possible); ecological damage associated with their "development;" clean-up of resulting pollution; social impacts; and every other consequence of their use. Because of appreciation and the effect of factoring in all future liabilities and expenses related to the development of resources, the NPVs of most proposed projects based on the conversion of resources into present wealth would be far into negative territory and deemed impractical.

Would the owner of an old growth tree think twice before cutting it to yield a few hundred dollars worth of saw timber if his

---

[199] From an article in *Business Week* by Philip L. Defliese, August 4, 1975, quoted by Barry Commoner in "The Poverty of Power: Energy and the Economic Crisis," Alfred A. Knopf, 1976.

[200] To express this thought another way, as quoted in *The Economist*, June 12, 2010: "Heard the joke about a businessman who asks his accountant what two plus two is? The accountant draws the blinds, leans over the desk and whispers: 'What would you like it to be?'" The article goes on to say, "the crafting of accounting rules is more art than science, thanks to the need to balance the interests of companies, their investors and—especially in banking—their regulators."

accountants assured him the tree will be worth thousands of dollars in a few decades? Would mining consortiums be formed to dig up bauxite reserves used to make aluminum soda and beer cans that end up in landfills if those reserves are shown to be of immense worth in the future?[201] The process of "developing" those resources destroys their capital value. But the imperative is to create present wealth, so the solution is to ignore all but the minimum effect on NPV.

If recognized and accounted for, the future capital value of resources would slow the tide of "progress." That cannot be allowed and thus this inconvenience is made to disappear through sleight of hand, merely by disclaiming or vastly underestimating the future value of resources. "Money," although often based on the negative value of debt, is assumed to increase ad infinitum, while the irreplaceable real capital of the planet is depleted for short-term gain and "profit." The chief acting force in this process is greed.

Economics Is Not a Science

Economists like to pretend that they are scientists of some kind. Nothing could be further from the truth, and therein we find the heart of this tragic tale.

True sciences are bounded by the laws of Nature, abiding by rules that are gleaned and applied from the reality of the universe. Economists in general have tended to accept no such boundaries, but have created their own "realities." In truth, they are more astrologers than astronomers, less chemists than alchemists. As we will see it comes down to the various definitions of the word "theory."

---

[201] It's likely that the landfills of today will become the "mines" of the future as our descendants seek to recover some of what we have squandered.

Before I penetrate too far into this intellectual swamp, let me make it clear that economists come in many kinds and flavors. There are literally dozens, perhaps even hundreds, of "schools" of economic thought. Some are more practical and enlightened than others, so to those I tip my hat and stipulate that my general criticisms must surely apply only to those of other schools. In defense of economists as a species it should be noted that it is often the creatures of politics and finance that make the biggest hash of things by favoring economic ideas that promise the Moon on a platter. Such strategies are appealing to politicians because fantasy and lies procure more votes than truth. Wall Street embraces them because they provide further licenses to generate wealth out of thin air.

Disclosure: I am not an economist, although I studied economics in college and have read and written about the subject throughout my life. The ideas in this book are based on those experiences, and on what I have always considered to be the "common sense" of economics, the kind responsible families use to budget their expenses.

There is an understandable tendency for the more "unreasonable" economic ideas to be put into practice, despite the disastrous results that sometimes result. Remember the lesson of an earlier chapter that it is the "unreasonable" men and women who are generally in charge of things, dedicated to the advancement of "progress." This is nowhere more true than in economic policy. Thus, it's the "unreasonable" economists—those who promise the most rainbows and unicorns—whose ideas are most often hammered into law or used to justify questionable financial practices. Economic models that call for sacrifice, in other words ones that take reality into account, are generally treated like a skunk at a picnic.

To understand the basic problems surrounding economics, I

suggest reading a little book[202] by Henry Hazlitt, who condenses the entirety of economics as it should be practiced into one single statement, the "lesson" of the book's title:

> The art of economics consists in looking not merely at the immediate but at the longer effects of any act or policy; it consists in tracing the consequences of that policy not merely for one group but for all groups.

In 26 brief chapters Hazlitt explores some of the economic fallacies that, as he puts it, "are working such dreadful harm in the world today [as] the result of ignoring this lesson." Through these examples he demonstrates the troubles with many popular economic policies. At the root of all is the failure to see (or deliberately ignoring) the longer-term effects of short-term policies, or the unintended (or deliberate) consequences borne by one group to the benefit of others.

True science works by posing theories that can then be tested against the real world. A "well-accepted" scientific theory is not a patchwork of speculation (although the original inspiration may have begun there) but is a model that has withstood a continuous process of rigorous testing.

The principle of reproducibility underlies all of science. A sound scientific theory can predict the outcome every single time. Those that fail to do so are left behind as science moves on in search of better models. That's the reason why such disciplines as physics and chemistry are called "hard" science.

For example, Galileo was testing the theory of gravity when he dropped cannonballs from the Leaning Tower of Pisa. Newton, Einstein and many others continued to advance the theory through

---

[202] H. Hazlitt, "Economics in One Lesson: The Shortest and Surest Way to Understand Basic Economics," Three Rivers Press, revised ed. 1979.

analysis and strict comparison with observed reality. The process is ongoing and there is much still to learn about gravity, but the essence of it is well understood.

When Galileo dropped his cannon balls, they always fell to the ground at the same speed and force, every time. The results were 100 percent predictable because gravity is a real force and brooks no violations.

Economists have no such fundamental internal process of connecting theory with reality, and all too often economic "theories" are based more on supposition than fact. Lacking reproducible experimental evidence, economic conclusions tend to be based on assumptions and most recently with the aid of computer models that are themselves skeins of suppositions. This process yields up a myriad of ideas from ingenious to questionable to wacky. Many such speculations are like castles in the clouds, coming and going almost with the changing of the seasons. None of them can ever attain the authenticity of scientific theory.

Unfortunately, in economics what seems to work at one time or place may yield completely different results the next time it's tried. Many if not most economic models may look promising under a certain set of conditions, but are likely to turn out to be wrong in the longer term or when the situation is different. When that happens, theorists merely turn to other assumptions, always on the hunt for the "new idea" that might win them a Nobel Prize or at least a position at a prestigious institution or government agency.[203]

---

[203] Like all academics in the publish-or-perish world of today, the careers of economists are determined by the number of papers they publish, thus encouraging a continuing flow of theories. According to an April 17, 2010 Economics Focus column in *The Economist*: "In 1988 Raymond Sauer of Clemson University calculated that a publication in a leading journal adds about

The art (let us call it that) of economics does not completely lack the ability to make predictions, but outcomes cannot be certain. Sometimes rosy predictions come true; sometimes, due to unforeseen factors, we have the Great Depression instead.

That's not to say that the practice of economics has no basis in reality. For example, the law of supply and demand can be derived in a general sort of way from observation. In normal circumstances (and there are various definitions for "normal"), this law predicts that when the supply of any good exceeds the demand for it, the price will go down because the vendors are bidding against themselves. When demand exceeds supply, the opposite occurs, because the buyers are bidding among themselves and thus causing the prices to rise.

But does this always give an accurate picture of economic events, a reproducible result? No, it cannot, because of the many factors involved. About a hundred years ago we may assume that the supply of buggy whips took a precipitous drop as people turned away from driving around in horse-drawn conveyances and whip makers closed their doors or began making tires, steering wheels or spark plugs. The supply of whips declined, but prices did not rise because demand had also disappeared.

Economists are okay with that, because in that instance supply was cut off by lack of demand, which was simply erased from the charts because potential consumers simply walked away. To make that work, the law of supply and demand further assumes that alternatives will always appear should supply and/or demand for a given item become redundant or exceed limits acceptable to the markets.

$3,600 [in today's prices] to an economist's salary. So over a young economist's career of, say, 30 years, each article makes them $108,000." The columnist suggests economists may publish papers because they're "only in it for the money." Good question.

A physicist could not be as sanguine should cannonballs begin to fall at varying speeds depending on the time of day, temperature, or positions of celestial bodies in the Zodiac.

In every instance such factors as changing consumer taste or need can throw a monkey wrench into the working of supply and demand. As long as there are alternatives, demand can switch, leaving supply as an irrelevant factor. There is little consumer demand today for black-and-white TV sets, wringer washers, 78-rpm record players, or cameras that require something called "film." The modern office has no use for carbon paper, typewriters, mechanical adding machines, or mimeographs. Like Elvis, demand has left the building, and supply has gone with it because there are better alternatives.204

The Error of Assuming Alternatives

Some forms of demand are quite flexible, exhibiting what economists call "elasticity." That is, if demand for a certain product or commodity is highly elastic, consumers can merely change their buying habits or easily go without. That concept works fine for widgets or other theoretical consumer items, but it's a different story when it comes to a basic necessity of life. The bottom-line example of this is food, which is an absolute requirement for human existence. Eating is not an option. The demand for food is about as elastic as a granite boulder.

And here's the rub: What happens when supply fails but there are no alternatives? When demand has nowhere else to turn in the face of falling supply? Here's what happens: Demand continues to exist and may even grow. The "haves" of the world will bid up

---

[204] This point was driven home to me recently when I went to an Office Max store in search of a bottle of ink for my admittedly old-fashioned (but elegant) Mont Blanc fountain pen. There was none to be found, and the young clerk looked at me funny when I inquired.

prices for the remaining supplies as long as they last, while the less fortunate go without. If the diminishing supply is of food, the result is famine.

The law of supply and demand is at work here, but what is being seen is a failure of the assumption that alternatives will always appear to replace any commodity or product that becomes over-priced or no longer appeals to the consuming public. This assumption is made by most schools of economics, for reasons related to the idea that growth must be without limits. For reality to be allowed to intrude into such economic models would be inconvenient for the continuation of "progress."

It's often assumed that not only will alternatives appear, they will be cheaper, better, and readily available in quantities to fully satisfy demand. This belief may have begun because it always seemed to work that way. The primitive Stone Age gave way successively to the Copper Age, the Bronze Age, the Iron Age, the Industrial Age, and at last the full blossoming of the High Tech Age that is now nearing an end. Those who judge the future by looking backward see an ever-flowering growth of "progress" and draw their conclusions accordingly.[205]

Inexplicably, but conveniently for the furtherance of "progress," the assumption of alternatives is even applied to non-renewable natural resources.[206]

---

[205] The examples of fallen civilizations are brushed aside as irrelevant, perhaps due to the assumption that those nation-states were filled with ignorant peoples and subject to poor management. It should be noted that many of those civilizations thrived for centuries and even millennia. Perhaps the economists of those times were not so ignorant after all, but in the end all civilizations must answer to Nature.

[206] It should be mentioned that so-called human capital has also been devalued, with a large proportion of the world population either unemployed or under-employed. Just as corporate entities despoil the Earth's natural resources, so do

As pointed out above, these assumptions make it possible to derive economic models that support the concepts of ever-advancing growth and progress. This is a world-view in which expanding GDP is considered the paramount measure of economic health, and any contraction deemed failure. Natural resources presumed to be essentially "free" are converted into ballooning quantities of goods, aided by the creation of credit, debt and "virtual wealth." This process powers the engine of industrialization down the dead end path we're traveling.

When it comes to natural resources, the assumption that alternatives will always present themselves is patently false. As we saw in chapter three, Nature makes no promise to yield alternatives to the natural capital on which our industrial civilization relies.

Common sense tells us that when it comes to irreplaceable natural resources, what we see is what we get and when it's gone there is no more. Rational analysis of this fact reveals that there will be an ultimate end to so-called "progress," and the troubles will travel up the chain from disappearing resources to manifest themselves as food shortages. As our industrial civilization nears the end of the dead end path, "progress" and ever-rising GDPs cannot continue as resources become scarce and eventually disappear altogether.

We've seen in earlier chapters how the depletion of natural resources is driving the planet toward scarcity of many of the things on which technological civilization depends. We are reaching or have already passed the peak of oil production, and yet demand continues to exist and there is no doubt that it would actually grow if it could.

they exploit human labor to their ends of attaining ever more profit and power. Industrialization results not in spreading individual freedom, but in poverty and hopelessness for the displaced.

There are energy alternatives to petroleum, of course, but for the most part they are less desirable, not more. They tend to be more expensive, less efficient, less convenient, less feasible (e.g. fusion power) and less sustainable (e.g., biofuels from crops). Many of the alternatives depend on "development" of the same natural resources that are already in decline (for example, through construction of wind generators, sophisticated energy grids and nuclear power stations, or mining the soil for biofuels).

Civilization is on a downward spiral of resource supply, both in quantity and quality. There appears to be no alternative that can compare with the products of petroleum, whether for fuel, plastics, fertilizer, or chemicals. The same can be said for many other natural resources that are subject to eventual depletion, such as iron, copper, fertile soil, fresh water and more.

Like members of the fictional crowd lined up to admire the Emperor's new clothes, many of the most popular economists fail to notice the plain fact that the eventual depletion of natural resources will bring their models crashing down. They ignore any future consequences of their programs by assuming that alternatives will always appear when demand calls for them. It's like stage magic: They wave a wand and the elephant disappears. Voila! And anyway, as Lord Keynes famously said, "In the long run we are all dead."[207] In other words, the future doesn't matter: we should focus only on the present and let the future take care of itself.

Unfortunately, the assumption that new sources of supply are always waiting just around the corner can no longer be relied upon as we move into a world of post-Peak resources. As noted in an

---

[207] J. M. Keynes, "A Tract on Monetary Reform," 1923. The full quotation is as follows: "The long run is a misleading guide to current affairs. In the long run we are all dead."

earlier chapter, Herbert Stein's Law[208] warns: "If something cannot go on forever, it will stop." Stein was one of the economists who "got it" about the ultimate end point of runaway exploitation.

Note his use of the keyword "forever," which points to the need for sustainability. Forever is a very long time, and even human history is a mere blink in the span of the universe and the planet Earth. If human beings expect to continue to exist as a species down through centuries and millennia of future time, they must agree to live within the limits imposed by the hard reality of the natural world, rather than the promises of economic prophets.

Needed: A Theory of "Real" Economics

The fact is that although economists are not scientists, we might imagine something we may call "true" or "natural" economics that is subject to the laws of the universe. A primary requirement is that any "true" economy be sustainable, because no long-term future can be imagined that does not require humanity to learn to live in balance with the natural yield of the planet. "True" economics cannot base its workings on unrealistic assumptions.

The fact that the human race is operating without the benefit of a theory of "real" economics poses a serious problem, because the hell-bent-for-leather technological world we live in is carrying us toward a yawning gulf that lies ahead at the end of the dead end path.

The most important point to take away from this is that the threshold of disaster is not in the distant future—it's upon us right now. In fact, the Oil Peak is almost certainly the key underlying cause of the recent economic crises that are rocking the world.

---

[208] Op cit.

Economists Can't Predict Human Nature

Besides the uncertainty of supply, another wild card that makes economic predictions less than certain lies in human nature. The vagaries of individual actions and inactions go far to understanding the disconnection between modern economics and the realities of the universe. This principle was described by the late financial advisor and political consultant Bernard Baruch, who wrote:[209]

> All economic movements, by their very nature, are motivated by crowd psychology. Graphs and business ratios are, of course, indispensable in our groping efforts to find dependable rules to guide us in our present world of alarms. Yet I never see a brilliant economic thesis expounding, as though they were geometrical theorems, the mathematics of price movements, that I do not recall Schiller's dictum: "Anyone taken as an individual, is tolerably sensible and reasonable—as a member of a crowd he at once becomes a blockhead." ... Without due recognition of crowd-thinking (which often seems crowd madness) our theories of economics leave much to be desired. It is a force wholly impalpable—perhaps little amenable to analysis and less to guidance—and yet, knowledge of it is necessary to right judgments on passing events.

Human beings are not like cannonballs, photons, or chemical reactions. It's because people can't be relied upon to act in a consistent way that economics in its present forms can never be "scientific." Repeatability is a prime directive of science, and peo-

---

[209] From Baruch's Foreword to a 1932 reissued edition of the classic book "Extraordinary Popular Delusions and the Madness of Crowds," by Charles Mackay, originally published in 1841. In this delightful book, Mackay described the process of mob thinking such as investment "bubbles" like the "Tulipmania" that swept over Holland in the 1600s.

ple are inherently unpredictable, whether as individuals or as groups. Psychology may play a more important role in the tides of commerce and industry than anything economists can contrive.

Even worse, humans respond to incentives and disincentives. Politicians and administrators recognize this, which is why they levy "sin taxes" on such undesirable practices as using cigarettes and alcohol. The supposed purpose is to discourage smoking and drinking, although the state is not shy about taking the money.[210]

Despite these inconvenient facts, many schools of modern economics have as basic tenets the assumptions that people will act in predictable ways and that every transaction is made with full knowledge not only of all the present factors involved but even of the future. This is difficult to reconcile with reality, since it implies omniscience. Anyone who can see the future has no need of economics and could soon become fabulously wealthy just by investing according to his or her visions.

Adam Smith, an important early economic philosopher and the originator of some of these ideas, wrote of an "invisible hand"[211] that guides the economic tides, thus creating the foundation of a flawed concept of self-regulating markets. According to his vision, every transaction is made with full understanding of

---

[210] As a side note, punitive taxes sometimes have the effect of aiding corporate interests, such as pharmaceutical companies that would much rather sell antidepressants than see potential customers bury their woes in a bottle or find peace in a quiet smoke. It's also interesting to note that when government becomes too greedy, individuals turn to so-called black markets or the "underground economy" to escape the consequences of regulation.

[211] A. Smith, "The Theory of Moral Sentiments," 1759. Smith's ideas formed the basis of the *laissez-faire* approach to economics, which prescribes markets and industry be unencumbered by government control. The world has seen the error of that idea time and again, most recently in the surge of bank and corporate failures that began in 2008. It has not stopped governments from letting markets run wild in the name of progress.

every factor by all parties involved. This is tantamount to evoking a magic wand to explain economic events.

One version of this idea used by economists today is the "rational expectations hypothesis," based on the assumption that everyone involved in the outcome of an economic model shares the modeler's view on how the economy works. Another is the "efficient markets hypothesis," which similarly assumes that prices always reflect all available information, in other words, are always rational. Yet another idea in this category is the so-called "representative agent," a kind of mythical "typical" actor in economic affairs, from whose actions the entirety of an economy can be derived.

These ideas don't stand up to the "common sense" test. If markets are always efficient, how can we explain panic selling and investment bubbles? How can we imagine that every participant in an economic transaction has full knowledge when it's well-known that buyers and sellers commonly exaggerate positive features and suppress negative ones? In a perfect world of rational expectations, no Ponzi scheme would ever be able to succeed. And because this representative agent is drawn from an economist's theories, rather than any concrete facts, how can he be substituted for every other player? It doesn't make sense.

All of these ideas sidestep the fact that humans have minds of their own and may sometimes, perhaps even often, act without a clear picture of their motivation. Not only do all minds not think alike, they're subject to change for myriad reasons, not all of which are rational or even sensible. The picture becomes even more muddled when we consider the effects on economic activity of manipulative psychological advertising, creating demand for products and services that no one in their right mind really needs. This may involve an invisible hand, but one engaged in overt manipulation.

David L. Brown

Pessimism and Realism

Part of the problem comes down to the different ways people respond to the situations in which they find themselves. We have words to describe those reactions, words such as "pessimism" and "realism."

Pessimism is a vague sort of idea, usually interpreted as a general suspicion that the future might be worse rather than better. Increasing knowledge of the facts brings pessimism into sharper focus. Its opposite, optimism, is often based on unrealistic expectations, denial or failure to understand the situation.

A "realist," according to one dictionary definition, is "a person who tends to view or represent things as they really are." They tend to be pragmatists, "somebody who only considers things as they are or appear to be, and avoids ideals and abstractions."[212] The opposite might be described as delusional, unwitting, or fantasizing.

Neither pessimism nor realism predisposes an individual to participate in the markets of everyday life as if guided by an invisible hand. That no one can know the future is an obvious fact that Smith and many of his successors chose to ignore. There is no accounting for the judgment of individuals who are paranoid, or acting from a state of ignorance. Those who are ill-informed or of a non-analytic nature and subject to a deluge of advertising and other psychological forces designed to warp their consuming practices are wandering through life without the aid of any invisible hand, hardly independent consumers but rather puppets or economic zombies.

Both pessimism and realism contain a measure of uncertainty.

---

[212] The Free Dictionary, Encarta.

Many of the realities that are presently sweeping across the globe, the results of traveling down the dead end path of industrialization, are becoming so clear, so indisputable to those with eyes to see, that they preclude nearly all doubt about the direction in which our planet is going. That leads to a quite different viewpoint about the state of humanity, namely that pessimistic and realistic mindsets are becoming more soundly based. It's getting a lot harder to be an optimist.

Both pessimism and realism can be intensified through the accumulation of facts. Realism might be seen as a more certain form of pessimism. In this view, the pessimist is one who is aware that some undefined danger may lie around every corner. The realist is one who has carefully examined the landscape and sees the tiger waiting.

To put the subtle difference between pessimism and realism into perspective, let's imagine a scenario in which a mountaineer is climbing on an icy and unstable rock face high on Mount Everest. The situation has become threatening and thus a high degree of caution is an appropriate response. At this stage of the adventure the climber might be regarded as pessimistic because he is worrying about the possibility of a mishap.

Now let's assume that the climber's pitons have pulled out of the rock, his companions have cut his rope in order to save themselves, and he is falling ten thousand feet to an icy glacier far below. Can we call him a pessimist now? No, for the facts are plain. There is no doubt and his future is certain, if short. Assuming he remains rational and sane, the only description that can apply to him now is "realist." Reality is about to smack him in the face with ten million tons of ice and granite.

The transition of the climber in our scenario from pessimist to realist is analogous to the transition that awaits our industrial civilization as it nears the end of the dead end path. Like the climber's reality of being in free-fall, the problems our civilization faces seem to be real, indisputable, and innumerable. Not only that, there appear to be few if any practical solutions to many of the social, economic and ecological threats we face.

It used to be said that every problem held within it an opportunity. That aphorism may no longer hold true in the post-growth world that lies ahead. The old economic model of "progress" as defined by never-ending growth is crashing against the reality of resource depletion, acerbated by over-population, greed, political and religious strife, and a multitude of other factors including, and far from least, the on-going collapse of the Earth's environment. All that is being facilitated by economic sleights of hand that dismiss reality and embrace fantasy.

Although few are yet ready to recognize it, the old world economic order is on intensive care. Those who pay attention are beginning to glimpse the new world that's beginning to emerge, and it isn't a pretty picture of puffy clouds, flowers, and butterflies. It's definitely not the world we've been promised, blessed with universal freedom, wealth and happiness made possible by an ever-growing global economy, through unending "progress."

The new economic order will enforce drastic change. Its imperative will be to create rational "true" economies that are resource-neutral and sustainable. To achieve that will require the almost total destruction of the old global economic and political systems—but never mind because they are in the process of destroying themselves.

Chapter Eleven

Entering the Age of Realism

We are beginning to glimpse what lies at the end of the dead end path. The old world order that is now coming to an end was built on the shaky foundation of what was presumed to be an ever-expanding supply of cheap energy and other natural resources required to keep the engines of commerce grinding away. It grew from a Utopian view that the world would always be a better, happier, richer, and more populous place—that "best of all possible worlds" satirized by Voltaire.[213] This despite ample evidence to the contrary that could be seen everywhere one might look, troubles of which we have been warned since the days of Malthus.

We are now witnessing the end of that road called "progress" down which humanity has traveled for millennia. It is the dead end path, and it leads to a breaking point, a sharp disconnect with the past. We are now in the early stages of that disconnection. Reality is coming into focus and it is no longer enough to be merely pes-

---

[213] In "Candide," published in 1759, the title character is indoctrinated in boundless optimism by his teacher Dr. Pangloss, only to encounter endless hardship and tragedy in the real world. Dr. Pangloss could be compared with many of today's economists and politicians.

simistic. We are in an in-between state, no longer quite in the metaphoric position of the mountaineer who is wisely cautious about his safety, and rapidly approaching the world-view of that somewhat later realist who is falling to certain doom.

Will the human race learn to embrace realism and seize the opportunities to make necessary changes, or merely hunker down in a state of denial? There is a battle raging over that question. Many forces are struggling to maintain the old ways (for their own continued benefit) without regard to the future. Still, the evidence is mounting everywhere that the planet is in trouble; that our industrial age is nearing a catastrophic end; and that the very existence of the human race may be in peril.

It's been a psychological roller coaster ride. The world outlook (at least in the advanced nations) has gone from a period of optimism that began with the Industrial Revolution and flourished during the economic boom following World War Two; proceeded into a period of vague uncertainty during the oil spikes and inflationary era of the 1970s and '80s; and began to morph into pessimism on the part of many of us after the tech stock market crash and 9/11 attacks. Now we are reaching the final stage where even pessimism is no longer quite adequate to set the terms of the situation we face. We have arrived at the threshold of the Age of Realism.

Those of us who have had our eyes open to the coming necessity for change have long been characterized as pessimists. Now, in the clear light of reality we no longer need to think of ourselves as somber old spreaders of gloom, always full of party-spoiling observations about the bad things that might be coming down the road, dismissed as conspiracy theorists (and everyone knows that no conspiracy theory can be true and that everything "just happens"). Today we are realists.

Unfortunately, there are too many others, particularly those in charge of things or who control far too much wealth, that have been slower to recognize, accept or admit the sharp turning point through which human history is beginning to pass. There are still too many mere pessimists in the world, and even a surprising number of ill-informed optimists.

Those many who remain unaware or refuse to admit that the old world is passing away are playing the role of the king's horses and men, trying to put back together an immense, planet-sized Humpty Dumpty, the representation of our economies, political systems, and societies that are beginning to shatter into a million pieces as we approach the terminus of the dead end path.

One of the signs of a system in decline or collapse is the evidence of denial. And indeed we are witnessing this to an almost shocking degree. Deniers proclaim that the future will be bright and wonderful, harkening to politicians who promise pie-in-the-sky and that "the best is yet to come." Prime examples of this are the concerted attacks against climate scientists in a coordinated effort to spread doubt about global warming, something virtually all legitimate scientists accept as real.

Simple comparison with the mounting evidence reveals this as denial on jet-powered roller skates. Even more disturbing is the lack of attention being paid to the coming industrial collapse. This phenomenon is entrenched all across the planet, where hardly any leader dares stand up and declare that an emergency is upon us. Instead, they continue to feed the beast, strive to keep the engines of "progress" working, and provide bread and circuses for the masses as civilization lurches toward almost certain disaster. Rather than facing the future and preparing us for it, we see our once and future leaders acting as would-be saviors of the old world order, pumping trillions of (perhaps imaginary) dollars into failed institutions in the apparent hope that if only those obsolete industries, entitlement programs, and other monstrosities can be put back on their feet, things will return to "normal."

Well, here's a hot tip: There is no "normal" any more. "Normal" fell out of the nest some while back and was replaced by its evil twin, "Abnormal".[214]

The idea that we can somehow transform a broken and failed system back into the "normal" world we once knew is an impossible dream, a vain hope, and a disaster in the making. To spend trillions trying to glue and patch and stitch Humpty back together is a foolish and pointless exercise in futility. We should be investing everything possible to make the future world work, not attempting to resuscitate the dying old one.

The Looming Specter of Famine

Nowhere does the economic trouble facing humanity loom more ominously than in the area of food security. There is growing concern about spreading malnutrition and famine among the third world poor. One response is to propose that rich nations spend more money to provide food for the hungry poor, but that ignores an important fact: The problem is not lack of money, but growing shortages of food.

This "solution" looks at the wrong end of the supply-demand equation, and understandably so because it's far easier to "create" money than food. All it takes is a printing press, and actually only a keyboard and a connection to the internet since most money is of the imaginary kind.

Even if more money could be thrown at the problem, it would

---

[214] It's fair to question whether the previous state was normal at all, or a historical aberration that Nature is beginning to correct. I vote for the latter.

not help, for the "haves" will always outbid the "have-nots" for scarce resources, and especially that most important commodity of all, food. The relatively well-off nations already have bid up the price of food, and to provide more money to poorer people would only put more bids on the table, causing prices to go even higher. The rich would double-down and the situation for the hungry poor would become even more ominous.

This is the basic principle of supply-and-demand at work, and we are reaching the point mentioned in the last chapter when no alternatives exist. For years—thanks to industrial agriculture, the Green Revolution, massive application of cheap energy, and "mining" existing resources of soil and water—food supplies remained high and prices low, keeping ahead of the rising numbers of consumers.

Now we are approaching what may be the most important resource peak of all, Peak Food. Even in its full-blown industrial form, agriculture is straining to increase production, pouring more non-renewable resources onto the land in the form of fertilizer and agrochemicals—yet demand continues to rise as long as population grows and emerging economies seek higher standards of living for their people.

Another predictable response to the threat of famine calls for more research in crop science, presenting the expectation that we can repeat the Green Revolution. Crop scientists ask for more money to fund their research, with the implication that if only we could spend more money there, the problems of food shortages would go away.

There may be a kernel of truth in this idea, as it was certainly true in the past when researchers were able to launch the Green Revolution. But that previous success rode on the back of abundant and cheap natural resources, and it took place in a world far less populous than today. It was not and never could have been a permanent solution to the problem of feeding humanity, because while human numbers continue to grow exponentially, the resources of the Earth are finite and are being rapidly depleted. As we learned in chapter two, this is a mathematical truth. Any future Green Revolution will need to produce far more from far less and at vastly greater cost. The possibilities are limited by the hard laws of Nature and cannot be modeled by economic theories based on convenient assumptions.

Law of Diminishing Returns

Another principle of economics, the Law of Diminishing Returns, assures that history cannot be expected to repeat; there can be no Green Revolution *Redux* because the game has changed.

The acronym "TANSTAAFL" represents a maxim that requires solutions to conform to reality and natural law. It stands for "There Ain't No Such Thing As A Free Lunch." It dictates that plans cannot be based on guesses, fantasies, or fervent prayer, and certainly not upon assumptions that don't take reality into account. It's a shame the principle of TANSTAAFL was embraced by engineers, not economists.

Unfortunately, the Green Revolution was a one-shot deal and any follow-up effort will of necessity be a quite different process, far more expensive, less productive, and plagued with uncertainties. It will not, as Yogi Berra put it, be *"deja vu* all over again." It's an unprecedented challenge. By recognizing the effect of resource depletion, realistic analysis tells us there is a high probability that agricultural science cannot succeed in preventing a reversal of population growth.[215] We are entering the slippery slope of the back side of a generic Hubbert curve of resources.

There are a lot of reasons why there can be no new Green Revolution, at least none comparable to the original. For one thing, the world population has approximately doubled since the Green Revolution began. For another, the world's resources of land, water, minerals and energy have been reduced and are becoming more expensive thanks to the law of supply and demand—and without suitable alternatives on the horizon. We've used up a huge proportion of those natural resources in creating the Green Revolution, and now we face an even more daunting challenge with fewer resources to bring to bear.

Ironically, the idea of diminishing returns was pioneered by, among others, Thomas Malthus, the economist and population theorist we met in chapter two. He and David Ricardo laid the groundwork for this concept, pointing out that fertile land was a limiting factor in agricultural production. As time went on and human numbers grew, the land would need to be used more intensively, causing yields to decline.

As we saw, Malthus was less concerned about catastrophe than he has been made out to be. His actual writings reveal that he believed that declining resources would act to keep human population in balance with Nature.

But the Law of Diminishing Returns has its roots right in the heart of the subject that is of serious concern today. Thanks to the

---

[215] In 2007 the United Nations stated that world food production will need to double by 2050 in order to sufficiently satisfy demand. That figure was later revised to "only" 70 percent, perhaps less on the basis of the facts than in recognition of the magnitude of the challenge. In view of resource peaks and depletion of soil, water, and other inputs, it's hard to imagine how such a goal could be met, seeing that food production already shows signs of faltering despite ever-greater use of fertilizer and other inputs.

use of industrial agriculture to force more production from the land, we face the long-denied danger of a "Malthusian catastrophe."

In fact, the Earth's soil is indeed being farmed more intensively, and in ways that Malthus and Ricardo could not have anticipated. As we saw in earlier chapters, the industrial techniques of farming include heavy use of fertilizer, herbicides, insecticides, irrigation, and powerful machinery—all made practical by the use of cheap and abundant petroleum.

And yet, even so, agricultural production appears to be reaching a plateau. It seems that the land can only be pushed so far. It was one thing to see corn production in Indiana climb from around 20 bushels per acre less than a century ago to 160 bushels now. That's an increase of eight times, or the effect of three doublings. Can we expect another doubling of corn production? Not very likely, and for many reasons, including the limitations of the land, of sunlight, of water, the resulting pollution of water resources through runoff, and even the possibility that the heavy applications of fertilizer required could be toxic to plant life. This is not to mention that future resources will not be available in quantities or at prices that would make such an effort feasible. It's all a result of the diminishing returns law.

"True" economics would predict that the most unfortunate of humans cannot afford food that is priced sky-high. Neither can rich economies even produce that food when the oil and other resources to do so are no longer available in sufficient amounts or affordable cost.

Alarmingly, the situation is more serious than even the realist Malthus could have foreseen. He knew nothing of the future science that would create the Green Revolution, nor did he know of the principle of overshoot-and-collapse; predict the coming of industrial agriculture; or anticipate that modern medicine and massive food aid programs would place population growth on fast forward. All of these factors have combined to cause world population to spike upward even as resources are dwindling.

As a kind of ironic example of devolved thinking over time, pioneering economists in the era of Malthus and Ricardo largely recognized that Nature must be taken into account. It was only later that theorists began to believe that economies could actually be managed, relying on such "magical" inventions as invisible hands, the creation of "wealth" from credit and debt, and the unfounded assumption that resources would continue forever to be in unlimited supply.

The diminishing returns law applies to almost every phase of the problems we face. For example, look at the subject of tillable land. For generations humans have broken up native prairies, cut down forests, and drained swamps to bring more land into agricultural production. The process is still going on in places like the Amazonian and Indonesian rain forests. At the same time, formerly productive land has been disappearing beneath housing tracts, highways, and parking lots, as well as through the process of erosion by wind and water.

As Malthus told us, there are hard and fast limits to the number of acres that can be brought into agricultural production, while unchecked human numbers increase exponentially. These are facts that any "true" economics must take into account.

A Gloomy Future for Humanity's Poorest

Put it all together and it's easy to see that as long as population continues to grow, science is unlikely to provide a satisfactory solution to the problems associated with a population that has outgrown the resources of the Earth. No matter what we

David L. Brown

might do to stimulate plants to grow faster and produce more, the limitations of soil, water, energy and everything else that Nature requires can only do so much.

It may be a harsh fact, but it is true nonetheless that food will never again be cheap or plentiful. That fact bodes very ill for the poorest humans who are already living on the edge of famine. The equations are simple:

Fewer Resources = Less Food
More People = More Demand
Less Food + More Demand = Higher Price
Higher Prices + Inelastic Demand = Shortfalls
Shortfalls = Famine.

At the end of the dead end path, when industrial agriculture can no longer be sustained and falls into collapse, only labor-intensive practices such as smallholding farms and cooperative gardens will be able to marginally contribute to the world's food supply. When resources are no longer available, "real" economics must prevail. No longer will huge tractors and combines move across thousands of acres of mono-culture cropland drenched in fertilizer and herbicide.

The need will be for more human labor and less reliance on mechanical power; more use of recycled waste and phasing out of chemical fertilizer; in short, more personal attention to the basic need of human beings to feed themselves. Future technology may help smooth the way out of the dead end path, but perhaps through such mundane contributions as the design of more efficient hoes and scythes, or of harnesses to get the most from animal power.

We are approaching an era in which there will be no easy

solutions such as have been tapped in the past, and which have had the questionable result of moving the human overshoot further ahead into dangerous regions. Now, the principles of TANSTAAFL and diminishing returns are coming into the fore and major change is inevitable, whether we want it or not.

The bitter truth is that there is no amount of wealth—whether used to buy food for the starving or in support of agricultural research—that will solve the problem. The overshoot is nearly complete, and the collapse is coming soon or already under way.

Reaping the Harvest of Inaction

There is a Chinese proverb that begins "The best time to plant a tree is twenty years ago." Much can be gleaned from this aphorism about the present state of the world food situation. The "solution" to the emerging problem, we are told, is to simply turn on the power of science and develop new crops that can feed the hungry masses. But to be of use today, that is a "tree" that must have been planted at some time in the past.

And the trouble is that agricultural scientists already have been laboring long and hard for decades to boost farm productivity. It's not as if the Green Revolution was achieved in a few short years in the 1950s and then all efforts were stopped. We have planted plenty of metaphorical trees and it's of no use to expect that we can start all over again as if there were a blank slate of opportunity to re-create the Green Revolution from scratch.

The Chinese saying concludes: "The second best time is now." Unfortunately, we must accept that for all the work that has been left undone, only the second best alternative is available to us, and if we fail to act in the present, only third best or fourth best choices will remain—or perhaps no choices at all.

In April, 2008, the cover of an edition of *The Economist* focused on the emerging world food crisis.[216] Unfortunately, the inside article was long on "crisis" and short on solutions. After reporting "some of the sharpest rises in food prices ever," the magazine ominously adds: "But this year the speed of change has accelerated. Since January, rice prices have soared 141%; the price of one variety of wheat shot up 25% in a day."

That's pretty strong stuff, but then the article goes on to state (emphasis added):

> The prices mainly reflect changes in demand—not problems of supply, such as harvest failure. The changes include the gentle upward pressure from people in China and India eating more grain and meat as they grow rich and the sudden, voracious appetites of western biofuels programmes, which convert cereals into fuel. This year the share of the maize (corn) crop going into ethanol in America has risen and the European Union is implementing its own biofuels targets. To make matters worse, more febrile behaviour seems to be influencing markets: export quotas by large grain producers, rumours of panic-buying by grain importers, money from hedge funds looking for new markets.

Soothing words indeed in the face of alarming rises in food prices. We are told it's not "problems of supply," but all due to the "gentle upward pressure" of demand, which merely reflects the higher expectations of a more fortunate world. A better world, indeed—perhaps even that "best of all possible worlds" of Dr. Pangloss. Food shortages are merely a sign of prosperity, you see, a good thing.

---

[216] "The silent tsunami," *The Economist*, April 17, 2008.

One might expect a magazine that calls itself after the very art of economics to know that supply and demand are linked like Siamese twins. When demand goes up one of three things must happen" 1) supply must rise to fulfill the increased demand; 2) prices rise until the excess demand is quenched; or 3) demand turn to available alternatives. If it were a simple matter of demand, and if supply-and-demand works as it's supposed to, new food production or alternatives would have quickly appeared and prices would have returned to acceptable levels. This is not what we observe.

It's an interesting fact, and certainly no coincidence, that petroleum prices were soaring at the same time as rice and other grains climbed to unexpected heights. Remember that food production is deeply dependent upon petroleum, and that oil supplies are peaking. That puts the focus squarely on the supply side, the result of input scarcity and skyrocketing cost. It is quite likely that the rising food prices in 2008 were a red flag indicating that a supply peak was being reached. Rising demand is merely an exacerbating factor in a developing crisis.

If you doubt the connection between oil and food, graphs from UNICEF show that changes in the price of wheat and the price of oil over the first decade of the 21st century track similar curves.[217] Graphs for other major foodstuffs show a similar relationship.

The truth is that due to resource depletion the world's ability to produce food is reaching a peak. The reason for those high prices is that food supply cannot be so easily or quickly increased, or perhaps not increased at all—and certainly not cheaply in the face of declining resources.

---

[217] "Aftershocks from the Global Food Crisis," a Social and Economic Policy Working Brief, UNICEF Policy and Practice, April, 2010.

The 2008 price rise on which The Economist reported can be clearly seen in charts of world rice prices, and it is striking. From a base in the range of under $4 to $6 per hundredweight in the first three years of the decade, the cost increased to more than $24 at the peak. But there's an even more important lesson to take from these statistics, and that is that even though rice prices pulled back from the sharp peak in 2008, the cost of rice appears to have settled in the range of $10 to $16, two to three times higher than just a few years earlier.

If as seems evident we have reached Peak Oil, this trend cannot be expected to reverse, and similar effects are seen in the case of other major food sources. The U.S. Dept. of Agriculture[218] has published its guesses about future prices of major U.S. field crops through 2020. Their projections going into the future are basically straight, with no indication of a decrease in prices. Oddly, they do not indicate the possibility of increases, either, although the ups and downs from 1990 to the present show anything but stable prices. As I write this in August, 2010, wheat prices have recently soared due to drought in Russia and other places and too much rain in Canada. The prices of corn, wheat and soybeans do not seem likely to drop back to previous levels prior to 2008.

We can presume that there are no price reductions ahead for rice, either. This poses a serious problem because rice is a primary foodstuff for many of the poorest people on Earth, those who live close to famine. Rice accounts for about 20 percent of all calories consumed by human beings, and 30 percent in Asia. We're dealing with human lives here, by the hundreds of millions. As staples such as rice rise in cost, the most vulnerable people in the world are

---

[218] From "USDA Agricultural Projections to 2019," a report released by the Interagency Agricultural Projections Committee, February, 2010. PDF downloaded 5/27/10.

being priced out of the market. Newly-rich consumers in China and other emerging economies are buying more food just as world production is struggling to keep up.[219]

And on the basis of what evidence does the USDA base its prediction that commodity prices will remain at approximately the current levels for the next ten years? In view of the facts we've reviewed in this essay that seems extremely unlikely. It smells a lot like wishful thinking or even disinformation meant to calm the masses and keep the world on the track of "progress," at least for a little while longer.

The dimensions of the problem are large, as indicated by this statement from an article[220] that appeared in *Science* Magazine in the summer of 2008:

> Over the past several years, more rice has been consumed than grown--the difference made up by dipping into world rice stockpiles, which peaked at 146.7 million tons in 2001 but declined to 73.2 million tons in 2006, according to USDA. Prices were already rising, then lackluster harvests, export restrictions, and speculative buying sent prices soaring. For example, a popular export variety of Thai rice jumped from $362 per ton last December to $1000 per ton in April. Prices have retreated to $720 per ton.

These statements mirror the message of the graph that appears above. Remember that it isn't the number of tons of grain produced that matters, but the per capita amount. Because population

---

[219] Not to forget the insanity of producing biofuels from farm crops.

[220] D. Normile, "Reinventing Rice to Feed the World," Science, 18 July 2008. On June 10, 2015, the price of premium Thai Hom Mall rice grade A stood at $1048 per ton.

continues to grow, a plateau in food production is a disaster in the making. When supplies begin to stagnate in the face of rising demand, growing numbers of people will face hunger and even starvation. In terms of "real" economics it's a certainty because we are entering a period when there are no alternatives to satisfy the demand for food.[221]

As we've seen, when the law of supply and demand cannot function as economists wish it to by "automatically" creating more supply as if by the action of an invisible hand, it becomes part of the dead end path down which industrial agriculture leads.

The *Economist* article[222] referred to earlier did recognize that serious difficulties are facing world food security. It noted: "Ideally, a big part of the supply response would come from the world's [450 million] smallholders in developing countries, people who farm just a few acres." It goes on to note the advantages of this approach, but concludes:

> Unfortunately, no smallholder bonanza is yet happening. In parts of east Africa, farmers are cutting back on the area planted, mostly because they cannot afford fertilizers (driven by oil, fertilizer prices have soared, too). This reaction is not universal. India is forecasting a record cereal harvest; South African planting is up 8% this year. Still, some anecdotal evidence, plus the general increase in food prices, suggests that smallholders are not responding enough. "In a perfect world," says a recent

---

[221] When famine sets in, we hear reports from places such as North Korea of starving people eating grass, but humans cannot process the nutrients in grass. In the cycle of nature, that is the role of the ruminant herbivores. Grass eating was also observed during the Irish Potato Famine, where the corpses of many victims were seen to have green stains around the mouth. There are no alternatives to "real" food.

[222] *The Economist*, op cit.

IFPRI report, "the response to higher prices is higher output. In the real world, however, this isn't always the case." Farming in emerging markets is riddled with market failures and does not react to price signals as other businesses do.

Well, there you go—economic theory says that small farmers should take up the slack to feed the world, but somehow they aren't doing their job. It's a ridiculous conclusion that the poor, the victims of exported technological agriculture, should now be expected to fix things. Again, the law of supply-and-demand fails to operate as it "should." The invisible hand is, well, not to be seen.

There's more to the problem than can be addressed by idealistic short-term "solutions." Increased demand is in part the result of expanding population, made possible by the era of cheap petroleum and industrial agriculture that is now nearing an end. Add in the effects of "progress," both in the rich and the developing worlds. And as the advance of "progress" introduces more people to higher standards of living, demand increases even more.

As we've seen, without cheap and abundant oil, there can be no cheap and abundant food. And unfortunately for them, about a third of the world's people have been able to exist only thanks to cheap and abundant food, much of it heavily subsidized by governments and shipped to the needy through food aid packages.

You may have noticed that pictures of food aid workers do not show them distributing pizzas, turkeys, or T-bone steaks. No, what you have seen are sacks of rice or flour that have been supporting the hungry poor of the world. It is for those staples that the poorest humans, many earning as little as a dollar a day, have been spending a large share of their income — or on which they rely from food aid.

Ponder the figure shown above for the price of rice and what it means to a poor family that has been barely hanging on with a subsistence diet. Just since the early 2000s, rice prices have increased by about 250 to 300 percent. That's not a minor blip, but a world-changing event.

What we are witnessing is the end of an era, an era of human expansion that's been accelerated in the last half-century or so through a combination of factors, including medical intervention, cheap food, rapid exploitation of resources, and a dollop of good fortune. The human race has drawn down the Earth's envir-onmental capital like a gambling addict with an unlimited credit line in Vegas.

To we fortunate residents of the rich world, food has been a relatively small part of the family budget, averaging in the range of 12-14 percent of expenditures in 2007 according to the U.S. Bureau of Labor Statistics. For those living in the "developing"[223] world, and especially the bottom third of humanity, food has been a much larger part of the family budget, perhaps 50 percent or more in many cases. For people in that position it's easy to imagine the disastrous impact of vastly higher food prices, the result of inelastic demand and faltering supply.

As we saw in an earlier chapter, foreign food aid programs have often been thinly disguised efforts by governments to reward the over-production of industrial farmers by bleeding off excess supply at the expense of taxpayers and thus artificially support

---

[223] I put the word in quote marks to indicate my opinion that it's no longer appropriate to apply the term "developing" to many parts of the world. The only "development" taking place, if any, is that carried out by rich corporations to take advantage of resources by destroying them. For most of the poorest nations, the term "failing" is a more appropriate description.

prices.[224] In an era of food shortages there will be no incentive for this, because over-production and the resulting potential for low prices will no longer be a "problem."

Further, should governments or international agencies seek to purchase food on the open market for distribution to the needy, in the face of supply shortages prices will be bid even higher in a spiraling process of inflation. It can be concluded that food aid programs are likely to shrink and even disappear as famine begins to spread in the "developing" world.

Some point out that higher food prices will benefit the poor by giving subsistence farmers more for what they grow. This is a false assumption. For example, the World Bank drew these conclusions (my emphasis):[225]

> In many poor countries, the recent increases in prices of staple foods raise the real incomes of those selling food, many of whom are relatively poor, while hurting net food consumers, many of whom are also relatively poor. The impacts on poverty will certainly be very diverse, but the average impact on poverty depends upon the balance between these two effects, and can only be determined by looking at real-world data. Results using household data for ten observations on nine low-income countries show that the short-run impacts of higher staple food prices on poverty differ considerably by commodity and by country, but, that poverty increases are much more frequent, and larger, than poverty

---

[224] We should not fail to note that American consumers take a double hit from this practice, first through taxation, second through elevated food prices. This is a prime example of the failure of economics to take into consideration the unintended consequences of economic policies.

[225] Summary of a World Bank paper titled "Implications of higher global food prices for poverty in low-income countries," April 1, 2008.

reductions. *The recent large increases in food prices appear likely to raise overall poverty in low income countries substantially.*

We can see that higher food prices likely produce more poverty, even though some among the poorest people of the world may enjoy higher prices for the food they produce. The reason is that there's a big Catch 22 involved. Yes, poor farmers could possibly profit by selling some of their crop at the higher prices, but the costs of inputs such as fertilizer, agrochemicals, fuel, and hybrid seed are also rising, and in fact those are important factors behind the higher food prices. As a result, many of the poorest farmers cannot afford even to plant the crops.

If they do succeed in harvesting something, they likely need to sell a large share of the crop to pay for the cost of inputs, and that could leave them with too little for themselves. The extra money will all have gone into the pockets of Saudi oil sheiks and mega-corporations, all acting in their own self-interest.

When there's no incentive to the small farmers, fields will be abandoned even as people go hungry. This goes far to explain the displacement of hundreds of millions from rural areas into overcrowded third world cities such as Jakarta, Manila, Mumbai, Mexico City, Karachi, Cairo, Lagos and Kinshasa.

One can imagine that the hopes of the industrial "haves" to sell technological wonders to an emerging new consumer class of former "have nots" will be dashed on the shores of reality. When those at the bottom of the economic pyramid cannot afford to grow crops or to buy sufficient food to maintain their families, how many TVs, autos, washing machines, cell phones, computers and cases of Coke are they likely to be in the market for? In a post Peak Food world, the poorest people will not only remain in the "have not" category, but many may join the legions of the "don't exist."

There is a basic economic fact that we should mention, and that is the difference between gross and net. If someone grosses a million dollars a year, and his expenses total a million and one dollars, the enterprise actually loses one dollar. He might brag to his friends that he has a million dollar business, but in fact he is a pauper who can't even afford a cup of coffee without a loan from the bank.

Whenever you see "progress" defined in terms of increased production or GDP, ask whether the gain was attained at a net loss. If so, it's not the wonderful thing the proponents of an ever-growing world economy want us to think. People who think in terms of gross are usually deluding themselves. It's what's left over that matters, and profit-motivated international corporations are proud to define their goals as to take every dollar they can from the markets they "serve."

Looking ahead to an era of declining resources and resulting higher costs for everything from oil to food itself, it's easy to see that more poverty will be the result. The dead end path leads to a downward spiral of economic factors that must inevitably end in widespread famine.

Entering an Era of Hunger

Now that we can no longer assume that plentiful and cheap food will be available, there are no obvious solutions at hand. The Green Revolution isn't the answer—we've already played that card. It's those past investments in agricultural science that have allowed us to get to where we are, with a world population of more than seven billion[226] and food production showing signs of peaking despite efforts by industrial agriculture to cash in on higher prices.

---

[226] As I finish preparing this second edition for putlication, the world population stands at 7,392,118,716, and rising at the rate of about five per second.

I have in front of me a book by Georg Borgstrom titled "The Hungry Planet"[227] that was published in 1965. Here is the first paragraph from the end paper blurb:

> The Hungry Planet is a chilling report on man's failure to live up to the simplest obligation of human survival: making sure of the next meal. Georg Borgstrom, a leading international authority on food utilization and world nutrition, points out that, even to feed the present world population adequately, we would have to double our food production overnight. With supreme ingenuity we might achieve and sustain this goal within forty or fifty years. But by that time, at the current rate of growth, our population will have doubled.

Well, fifty years from the publication date is now in 2015 and Borgstrom's predictions can be examined with the eye of a Monday morning quarterback. Here are some relevant statistics:

• World grain production in 1965 was 905 million tons. Production in 2015 was an estimated 2.527 million tons. In the early part of the 40-50 year period, grain production reached 1,984 million tons in 2006, according to USDA estimates. In that year, production had fallen short of demand in six of the previous seven years. Borgstrom's suggested target of doubling agricultural output was achieved through Green Revolution actions.

• Borgstrom estimated there were 3.25 billion humans on the Earth at the time he wrote his book. A doubling of that would indicate a population of 6.5 billion. In fact, at the end of 2015 there were more than 7.3 billion. If anything, Borgstrom's dire prediction was actually on the low side, it seems.

---

[227] Georg Borgstrom, *The Hungry Planet: The Modern World at the Edge of Famine*, The MacMillan Company, 1965.

So where do we go from here? *The Economist* certainly is partly right that higher food prices are due to increased per-capita demand. That's part of the picture, but doesn't change the facts for those in the poorest third of the human population. As we saw in an earlier chapter, in order for food supplies to keep up with a doubling of population growth while at the same time satisfying rising expectations for better standards of living, they must actually be tripled, not merely doubled. In the 50 years since Borgstrom's book appeared, supplies have a bit more than doubled, not tripled.

After a half-century of Green Revolution "success," the world is indisputably worse off, not better.

There is a wide disparity between those at the top of the human "food chain" and the one-third that are already living with malnutrition. This is a social problem that causes much anger and hand-wringing, but as usual there are no easy solutions. Can we suppose that the more fortunate classes in rich and emerging nations are going to willingly give up their higher standards of living to accommodate the growing multitudes of the poor? Probably not. For a nouveau riche Chinese to improve his family's dietary standard of living, some less fortunate family may starve. For that matter, is it likely that the French will give up their cheese and escargot, the Japanese their sushi, the Germans their strudel and wurst, or the Italians their pasta so that others may not starve?

And what about here in America? How many of us are prepared to go hungry two days a week so that others less well off may not starve? Well, some of us may make a few gestures,[228] but

---

[228] The difficulties of bridging the food gap are immense. I am reminded of a story my wife tells from when she was very young. At dinner one night, her father told her to "eat your food or some little Chinese girl will starve." My wife's serious but naive response was to push the plate to him and say "Well, send it to her." Would that it were so easy.

the will just isn't there for those of us in the rich world, and espe-cially in the newly emerging economies, to make major sacrifices. Sure, we could donate some money to help—but as we've seen, money won't buy food that isn't there. The new reality is that it's a rock-and-a-hard-place for the bottom third of humanity.

Unfortunately, we're at that point where the equation between human population and Earth's natural resources cannot be pushed much further on the resource side. The human "footprint" on the environment already requires "more than one Earth." Something else must give, and it will almost certainly be human numbers that must adjust. Unfortunately, that adjustment will probably take place in the shorter term, rather than through natural attrition over time.

Let's return to the proverb I quoted earlier: "The best time to plant a tree is twenty years ago. The second best time is now." Well, in the case of the food crises that are beginning to unfold, "now" is just entirely too little, too late. We've already done most of what could be done. The Green Revolution succeeded only in doubling human numbers and cannot be repeated.

We've examined the Peak Oil predictions of M. King Hubbert, the petroleum geologist who devised the famous curve that graphed the rise and fall of petroleum resources. His projections for U.S. production proved spot-on, and the world peak is being reached now or may already have been passed.

Now it's interesting to note that almost any finite resource (and there is no such thing as an infinite resource) can be tracked in much the same manner. The resource will be utilized until the laws of supply-and-demand and diminishing returns drive supplies to a peak, after which continued depletion will take place. Supplies will fall and demand will force costs ever higher, pricing more and more people and nations out of the game.

It could be argued that the Hubbert analysis doesn't apply to food, which is after all a renewable resource. But take into account the fact that industrial agriculture is heavily dependent upon non-renewable resources and we can see that Food Peaks are quite easily predicted by the bell curve. Remember the simple equation: Oil = Food.

We've already seen peaks in the case of ocean fisheries, most of which are sliding down the backside of the depletion curve. Production of staple grains is teetering at the peak even as population continues to rise. Peak Food is fast approaching or even already here.

And what does that mean? It means that we can imagine a new "Hubbert curve," one that graphs population. Picture the Hubbert Peak Oil graph and imagine that instead of oil, that rising, then falling curve represents numbers of human beings. Besides Peak Oil, we can now begin to contemplate the approaching event that could be called Peak People. It's an inevitable result of the way things work in the hard and cruel real world, due to realities that too many economists conveniently overlook in their desire to promote "progress."

What might this event called Peak People look like, and how does it relate to Peak Oil? A graph posted on a leading oil resources web site attempts to combine the two related factors and predict the future drop in human numbers.[229] Population is represented as a Hubbert bell curve that peaks 15 years or so after oil and follows a gentle downward path. The reality is the back side of Peak People is likely to be more precipitous, and to follow more quickly on the fall-off of oil and oil substitutes.

---

[229] Source: Tony Eriksen, "The Oil Drum" www.theoildrum.com. Eriksen, a regular contributor to this site under the handle "ace," holds a degree in engineering and an MBA. Graphic posted May 20, 2008.

David L. Brown

As we saw in earlier chapters, overshoot-and-collapse is a demonstrated fact of Nature. It has caused the failure of past civilizations, but never before has it occurred on a global basis. As a result of the continued rise of human numbers beyond what the Earth can support, overshoot has occurred. Now we are glimpsing the beginning of the collapse that must surely follow in its wake, as sure as thunder follows lightning.

Chapter Twelve

Toward a New Economics

As we've seen, the problem with some economic theories is
that they do not take reality into account. There are exceptions
among economic thinkers. For example, there are trends underway
to new kinds of economics that factor in "natural capital," the real
value of the environment, biological diversity, and other forms of
"wealth" ignored by modern schools of economics. Such an
approach accounts for the actual value of natural resources and
"ecosystem services." According to their proponents, such
approaches can yield highly efficient and profitable business
models; however this is yet to be demonstrated.

One such theory is set forth by Paul Hawken and Amory and
L. Hunter Lovins who call their model "natural capitalism."[230] The
authors make many of the same points we've covered above,
including the fact that economic theory "works" only by making
assumptions that cannot be supported by reality. Discussing the
way greed has ramped up in recent decades, they write:

---

[230] P. Hawken, A. and L. H. Lovins, "Natural Capitalism: Creating the Next
Industrial Revolution," Little, Brown and Company, 1999.

The 1980s extolled a selfish attitude that counted only what was countable, not what really counted. It treated such values as life, liberty, and the pursuit of happiness as if they could be bought, sold, and banked at interest. Because neoclassical economics is concerned only with efficiency, not with equity, it fostered an attitude that treated social justice as a frill, fairness as *passé*, and the risks of creating a permanent underclass as a market opportunity for security guards and gated "communities." Its obsession with satisfying nonmaterial needs by material means revealed the basic differences, even contradictions, between the creation of wealth, the accumulation of money, and the improvement of human beings.

Another concept is called "Green GDP," through which the damage and loss to the environment is factored into national scorecards. Ominously, the first attempt to apply this method, by China in 2004, yielded disastrous results that were "politically unacceptable." According to a Wikipedia article:[231]

"In the face of mounting evidence that environmental damage and resource depletion was far more costly than anticipated, the government withdrew its support for the Green GDP methodology and suppressed the 2005 report…"

This example illustrates the difficulty of moving toward a "new" economics. Green GDP failed because the Chinese could not let go of the idea of growth as the ultimate measure of economic success. Similar difficulties would surely be encountered anywhere such a model might be considered.

Any "true" economics must recognize that the highest sustain-

---

[231] "Green Gross Domestic Product," Wikipedia, downloaded 5/21/10.

able level is one of no-growth, and to get to the point where even that would function will require an extended period of negative growth to compensate for the excesses of the past. In other words, to get to the distant plain, we will first have to cross a deep canyon.

Another voice for a new kind of economics is Lester R. Brown, founder of Earth Policy Institute and author of many books about the environment, agriculture, and the fate of humanity. He calls for what he calls "eco-economics," and predicts that we need a "Manhattan Project" kind of effort to pull us back from the brink.

As with any rational plan, a stable or even declining world population is a cornerstone of Brown's proposals. He makes the point that economists generally have no training in ecology, and thus make poor decisions. In his book *Eco-Economy*,[232] he sums up the situation thus:

> Today's global economy has been shaped by market forces, not by the principles of ecology. Unfortunately, by failing to reflect the full costs of goods and services, the market provides misleading information to economic decision makers at all levels. This has created a distorted economy that is out of sync with the earth's eco-system—an economy that is destroying its natural support systems.

Brown's statement is a fair summary of the subject of this chapter. Further addressing the effects of the distorted economy, he writes:

> An economy that is in sync with the earth's ecosystem will contrast profoundly with the polluting, disruptive, and ultimately self-destructing economy of today—the

---

[232] L. R. Brown, *Eco-Economy*, W. W. Norton & Company, Inc., 2001.

fossil-fuel-based, automobile-centered, throwaway economy. One of the attractions of the western economic
model is that it has raised living standards for one fifth of
humanity to a level that our ancestors could not have
dreamed of, providing a remarkably diverse diet,
unprecedented levels of material consumption, and
unimagined physical mobility. But unfortunately it will
not work over the long term even for the affluent one
fifth, much less for the entire world.

Another advocate of the idea that economics needs a complete
redesign is John Ikerd, a professor emeritus of agricultural economics at the University of Missouri. In one of his books[233] he writes
(with heavy irony):

In economics, by definition, human wants are insatiable
and resources are scarce. Thus, we can never have
'enough.' The more we get from our scarce resources,
however, the better off we will be. So the goal of
economics is to make as much money as possible so the
economy will grow as fast as possible and become as
large as possible.

That, in a nutshell, is the principle behind the concept of
"progress," the endless drive for "more, More, MORE" that lies
behind the surge of industrialism that has driven us down the dead
end path.

Is it reasonable to assume that we can never have enough? In
America we have seen the definition of "poverty" constantly raised,
again and again, until many Americans labeled as "poor" are
surrounded by what previous generations would have considered
wealth beyond imagining.

---

[233] J. Ikerd, *A Return to Common Sense*, R. T. Edwards, Inc, 2007. See the
Bibliography for other books by Dr. Ikerd.

In 2015, according to the U.S. Census Bureau, a couple under the age of 65 earning less than $15,930 was considered to be living in poverty. But it's relative to the situation. If the couple were living in Manhattan, New York, that would likely be true. If living in Manhattan, Nevada (population 124) it would be less so. While truly poor families in some parts of the world live on a dollar a day, for an American couple to be counted as poor in 2015 required them to have income of $43.64 a day or less.

To repeat, it's all relative. A "poor" American couple living on $15,930 a year in a tax-friendly rural location with a modest paid-off house, driving a wholly-owned and well-maintained older car, storing food from a large garden and selling a few fresh vegetables in season, and managing a chicken coop for eggs and meat might very likely be far better off in real lifestyle terms than a couple earning $100,000-a-year and living in a Manhattan Island apartment.

Ikerd considers the rise of international mega-corporations to be a root cause of the problem. He writes:[234]

> Most economists today have no idea of the raw economic and political power of today's giant corporations. Competition is no longer "workable"—it simply doesn't exist, at least in an economic sense. Entry of successful new firms into established industries is virtually impossible, as is the orderly exit of firms. Advertising has evolved from persuasion to brain-washing, and consumer sovereignty has been replaced by corporate sovereignty. Capitalism has been replaced by corporatism. There is no theoretical foundation for the corporate economy in any economics textbook. Econo-mists simply deal with the economy "as if it were"

---

[234] Ibid.

competitive capitalism. Clearly, it is no longer either competitive or capitalistic.

Corporations exist essentially for two simple goals: to profit, and to grow larger so that even greater profit can be amassed in the future. To a "well-managed" corporation, people are nothing more than numbers on spreadsheets. Top executives almost always include accountants and lawyers, the former to pinch every penny of profit, the latter to advise on how close the corporation can skirt the limits of the law in seeking those profits.

Through support of politicians (both reported and otherwise), corporations have seized effective control of governments around the globe, not least in the United States where in the wake of the ongoing economic crisis that began in 2008 the taxpayers were saddled with trillions of dollars to bail out such "too big to fail" behemoths as General Motors and insurance giant AIG, not to mention almost the entire top end of the banking industry.[235] Even as the bailout took place, in the aftermath of the greatest financial disaster since the Great Depression, Wall Street bankers received record bonuses in 2009. The big got bigger, the rich got richer, the powerful gathered even more power.

The Truth About Money

Clearly, our present methods of economic calculus are not working except to the benefit of politicians, financial moguls, and corporate chiefs, all of whom "game the system" to fatten their own wallets and extend the reach of corporations over what they define as "wealth," that is to say, "money," for these modern-day buccaneers all-too-often recognize monetary units above all other

---

[235] Strangely, smaller correspondent banks were left to fail and be swooped up like spoils of war by larger banks operating on taxpayer life-support. Where's the sense or moral justice in that?

values, including those of basic human decency, fairness, and honesty.

Money once had innate value. For example, gold or silver coins were made of precious metals, and paper currencies could once be redeemed for those same metals. Until a few decades ago, dollar bills were called "silver certificates," for they served only as a proxy for the metal itself. Money was actually "worth" something, because it was tied to actual things such as the gold in Fort Knox.

Prior to the Bretton Woods international monetary system established at the end of World War Two, the relative value of currencies was based on a fixed price for gold. By raising that price, governments could "cheat" on their debts by devaluing the currency. But there were limits to how far that scam could take us rushing down the path of "progress," so currencies were steadily disconnected with underlying assets such as gold, silver and platinum. In 1972 the price of gold was allowed to float, no longer tied to the value of the American dollar in any way.

Today the relative values of currencies are based on each other (e.g., the Dollar vs. Euro, Dollar vs. Yen, etc.), none of which are asset-based. To a significant extent the values of the dollar and other currencies have become meaningless statistics. And when the inconvenient need to back their currencies with assets, govern‑ments can "cheat" on their debts merely by turning on the printing presses. As long as "everyone's doing it," the effects do not become obvious.[236]

---

[236] It seems self-evident that although currency A might remain relatively stable by comparison to currency B, that does not imply that either is actually on solid ground. The relationship could be the same if both currencies are falling at a nearly equal rate. In fact, if most of the world's currencies are tied together like

Today's "money" is a phantom, something that exists in computer networks and can flit around the globe at the press of a key. It's a fiction, a mere cloud of bits and bytes. Its only value lies in the ability of people and institutions to "believe" in it. It's fairly amazing that so many have fallen for this sham, like the cult followers of some charismatic leader. But then, what options do we have? Few of us can exist in the underground economy of barter. Meanwhile, the real assets of the Earth, its irreplaceable natural resources, are being destroyed to feed the hungry maw of consumerism.

This disconnection of money with real wealth is particularly notable when we speak of paper assets that have only a secondary or tertiary relationship with the underlying value, such as futures options, naked short-sales, derivatives and the like. The value of these things may not only be based more on faith than fact—they also require more than a dollop of credulity and fantasy, akin to the belief in pots of gold at the ends of rainbows. To a very real extent, the kind of "money" they represent just doesn't exist in any real sense because it's based in large part on supposition, hope, and, yes, faith.

But what about "real" money in the form of actual paper bills or coins? Surely a dollar is worth, well, a dollar. And that's true, as far as it goes. But if you are familiar with the concepts of inflation and deflation you know that the "value" of the dollar can rise or fall. That means that at any given time a dollar will buy more or less of any given commodity based on the fluctuation in its value. Stash dollar bills in your mattress and the future worth will depend upon the trends of inflation or deflation, not on any value innate in

---

the strands in a cat's cradle, what you have could be described as a tangled mess.

the cash itself.[237] And, of course, there's always the possibility that the issuer could collapse. How much was a Confederate dollar worth the day after Lee's surrender at Appomattox?

Should one hoard hard assets instead of cash, gold ingots for instance? A good idea, right? Sorry, but the value of gold and any other commodity also fluctuates, rising and falling with demand and supply. There is real value to metals such as gold, but their present values can vary.[238] Although gold has recently stood at around $1200 per oz., as recently as 2002 it's been below $300. And just try to buy a sack of potatoes with a gold ingot—it just isn't functional as a medium of exchange.

Now let's consider another fact that helps illustrate the "fantasy and fiction" nature of money. Ask yourself how much you "lose" when your investments drop. Sounds simple, but it depends on a lot of factors.

Let's say that on a hot tip from a taxi driver who recently arrived from Waziristan you bought 1000 shares of stock in Widgets-R-Us when it was selling for $5 a share. That's an investment of $5000. Let's assume that during a moment of financial euphoria the price of Widgets-R-Us shares soared to $70. Your holding was then worth $70,000 and you were "richer" by $65,000. Whoopee!

---

[237] The value of present-day coins, particularly copper-plated zinc pennies, is actually real to some degree. It costs the U.S. Mint about two cents to manufacture a one-cent piece. (Don't ask why this makes sense to anyone.) However, it would be impractical to conduct one's financial affairs completely in small-denominational coinage.

[238] Gold and other commodities can be a hedge against the value of cash, since a falling dollar has historically caused the value of gold and other precious metals to rise, although even that rule has begun to break down.

But wait, now comes some bad news for widget futures and Widgets-R-Us stock drops by half and is now worth only $35 a share. You sell and complain that you just lost $35,000. But did you really? Well, no. In actual fact you still have $35,000 left, and your original investment was only $5,000 so you have made a 600 percent profit on your investment. That other $35,000 that you think you lost never existed except as an unrealized potential, a phantom.

See how slippery those "dollars" are? Sometimes they can be seen as real, sometimes they seem to disappear like the morning mist. But the secret is that it's a casino game from the start. Whatever you gained or lost was just on paper and did not exist until you cashed out and walked away from the table. Meanwhile, you were just letting your "winnings" ride, as Las Vegas gamblers sometimes do.

Now if the value of money can't be relied upon, if money is indeed a fiction and a phantom, then what can we conclude from that? Well, for one thing we should recognize that there are certain values that are innate in the economy. For example, while currencies might crash in value as has recently occurred in Zimbabwe, where a 250 million Zimbabwean dollar bill[239] might buy you a loaf of bread today, hard assets such as real estate retain core value as long as you step back and stop thinking of them as chips to be used in a casino.

For example your house should be viewed as the place where you live, a cost of living, not a speculative investment. A big part of the problem with the real estate market today is that when home values were rising, too many people gave in to the temptation to "refinance" as their house values rose, taking out the capital

---

[239] After printing bills in denominations as high as 100 trillion Zimbabwean dollars, in 2015 the country adopted the U.S. dollar as their official currency.

appreciation and spending it instead of letting it build up over their lifetimes.[240] They are encouraged in this by institutions that seek initiation fees and new mortgages to consolidate into over-rated derivative packages for sale to unwary investors. They are further tempted by a constant barrage of advertising urging them to buy the latest model car, the biggest TV, or the newest computer game. It's another example of substituting debt for wealth.

When seen as a long-term investment for the purpose of providing shelter, a house was once a fairly rational form of capital. It's only when the real estate market is turned into a casino that trouble arises. If you are in it for the long haul it shouldn't matter to you whether you could sell your house high, low or in between. Live there until you die and it won't matter how much it's worth at any given time.

But even that is a questionable benefit, because you can never really "own" your house due to the power of government to tax it. A tragic fact of our present broken system of economics is that that many people, particularly the elderly, own their homes outright but cannot afford to live in them because the tax burden has become too great. In effect, the system as it now exists allows us to purchase our houses at great profit to corporate entities and financial institutions, and then "rent" them from the government through payment of real estate taxes. Again, true capital has been transformed into "virtual wealth," all to the benefit of corporations, financial institutions and governments.

---

[240] This is a kind of reverse form of the investment strategy known as dollar cost averaging, in which one buys a fixed number of dollars worth of a stock or bond at regular intervals. This is effective because when the market value is low more shares are purchased, thus increasing the portfolio's worth over the long term. In the case of house refinancing, the effect is the opposite and the "investor" ends up with less equity over time because the increases are drawn out and squandered.

The same logic once applied to the markets for stocks and bonds, which were purchased by prudent investors for the long term, even from generation to generation. They represented real value, at least to some extent (like the Confederacy, corporations sometimes disappear along with their value). This was the original form of capitalism, the one that Dr. Ikerd tells us no longer exists.

Here's another example of the conversion of capital markets into games of chance. Futures options were created to allow hedging, so that a large bakery, for example, could lock in the future cost of wheat. These instruments were not intended for trade except between the suppliers (farmers and grain dealers in this case) and the end user. The purpose was to stabilize the market process, not create a game of chance for speculators. Speculation undermines the purpose of any system intended to smooth out a market, to the detriment of the intended users of that system and the unwarranted benefit of the speculators.

Another grievous example is the use of credit derivatives, basically a straightforward side-bet against a particular bond or other financial instrument. Originally designed as a form of insurance to protect investors, derivatives quickly were seized upon by speculators. For example, for a cost of about one percent per year of its face value, a "wager" could be placed that a certain bond would default. If it did the "investor" would receive the full value of the bond even without owning it, thus obtaining a 100-fold return. Many billions were made in this way during the 2008 financial crisis, in many cases using borrowed money and thus once again showing how debt can be converted into wealth. Examples of this type of instrument include credit default swaps, collateralized debt obligations, and a dozen or so other forms.

Although they may originally have been created to serve serious purposes these markets have turned into a kind of global casino that puts Las Vegas to shame. Day traders, hedge fund managers, monetary fund speculators, and a host of other professional gamblers have turned the financial world into what amounts to a circus sideshow of whirling roulette wheels, craps tables and slot machines grinding away 24/7/365. The world of finance has been taken over by gambling addicts. On-line trading has replaced the bookies and numbers runners of a former era and spread the gambling bug far and wide.

And—it's legal! No worries there, because politicians always are well taken care of by financial institutions from Wall Street and around the world. In many regards, the world of investing has turned into a giant Ponzi scheme—just ask Bernie Madoff—and as in Vegas, those who truly control the system always win.

Just imagine all that "money" to be made, piling up in accounts from Wall Street to Lichtenstein to the Bahamas, and flowing in the form of electrons and microwaves hither and yon all around the globe. Trillions of dollars worth of "money," and yet … what is it really, and when markets collapse, where does it go? Here's where it goes: to the place where fairies lurk, where leprechauns hide their pots of gold, the place where dreams go when the dreamer awakens. Those things never existed, any more so than unrealized paper profits are "real money."

David L. Brown

Chapter Thirteen

Economics on the Farm

As I made clear at the beginning of this extended essay, the problems resulting from industrial agriculture should in no way be laid at the feet of farmers themselves. In fact, farmers as a whole have been primary victims of industrialization. Those who have survived thus far may appear to have gone willingly down the dead end path. This perception ignores the fact that they have had little choice in the matter, and that the process of farm industrialization imposed by corporations with the collusion of governments has destroyed the livelihoods and futures of hundreds of millions of people around the world.

We've touched briefly on the effect of agricultural corporatism on subsistence farmers in "developing" nations. No longer able to afford the tools of technological farming—machinery, fertilizer, chemicals, hybrid seeds and other "gifts" from corporate agribusiness—hundreds of millions of smallholders have abandoned the countryside and moved to slum-infested cities in search of often-illusive employment.

Rich countries further load the dice against farmers in poor nations through tariffs and price supports that protect their own big farmers at the expense of smallholders. The NAFTA agreement, for example, put tens of thousands of Mexican farmers out of work and that country's tortillas are now made largely from cheap U.S. corn, the product of industrial agriculture.

According to a recent paper[241] this is a result of U.S. subsidies, concluding: "it is evident that by lowering both the domestic price of corn and employment levels of corn farmers, federal subsidies for American corn are primarily responsible for the post-NAFTA rural to urban migration of Mexico's corn farmers." As evidence, the author noted that between 1995 and 2006 American corn growers received more than $56 billion in subsidies.

"Rich" Farmers Are Victims, Too

Despite the evidence of high subsidies, the process of industrialization has brought harm to farmers in the rich world, too. In the United States, farmers have been under constant pressure by government, agribusiness, and lenders to "get bigger or get out." In other words, either take over your neighbor's land, or give your land over to him.

Why might this be? It's a natural extension of the ideas of "progress," the endless drive for growth. It's something that works for the benefit of the rich and powerful, not ordinary folk and certainly not farmers who are caught up in this whirlwind of change. As we've seen, GDP growth is viewed as the be-all and end-all of economic success, and whenever change is taking place

---

[241] Rick Relinger, "NAFTA and U.S. Corn Subsidies: Explaining the Displacement of Mexico's Corn Farmers," *Prospect: Journal of International Affairs at UCSD*, April, 2010, University of California-San Diego.

there's always a buck to be made, by real estate agents, loan officers, agribusiness dealers, insurance salesmen, and everyone else who "serves" the farmer.[242]

Corporations find it more profitable to deal with fewer and larger farm operations. The process of consolidation boosts their opportunities, too. An environment of constant change creates the need for never-ending investment. For example, when a farmer takes on more acres he or she almost always needs to assume more debt, not only to buy the land but to acquire more of everything industrial agriculture demands. Agribusinesses, lenders, food processors, marketers—all these benefit from the endless cycle of "progress," while the costs and risks involved are laid solely onto the back of the farmer.

Financial institutions encourage the "bigger is better" model because it results in the expansion of credit, the commodity on which they thrive. Farm lenders are always eager for their clients to take out new loans to buy additional land, a larger diesel tractor, or the latest genetically engineered seed at several hundred dollars a bag.

Needless to say, land and other assets are held as collateral and seized when hard times come and farmers are unable to keep up with the ever-growing mountain of debt they took on because they were told they "must" in order to compete.

Industrial farming requires a huge amount of cash, and few farmers are in a position to finance expansion and on-going operations without the use of credit. Farm loans are a huge source

---

[242] I'm reminded of a 1950 science fiction story by Damon Knight in which space aliens arrive on Earth apparently prepared to aid humankind. Their handbook, "To Serve Man," is discovered to be a cookbook. The story was the basis for a Twilight Zone episode.

of revenue for lenders, and principal and interest payments are a major expense item for farmers. The use of credit protects everyone except the farmers themselves, allowing suppliers to be paid up-front and without risk.

Politicians, too, would rather pander to a consolidated farm economy, handing out largesse and accepting votes and contributions from a smaller base of rural citizens, and thus empowering the cities. Through this process farmers have been denied their proper voice in the affairs of a nation that depends upon them for its basic survival. This is ironic in view of the fact that at the time our nation was founded about 95 percent of Americans were farmers and planters, including many of the Founding Fathers such as Washington, Adams and Jefferson.

Government agencies grease the wheels of progress by providing educational support and management advice to farmers through federal, state and county agricultural extension agents, experimental programs at universities, publication of pamphlets, training programs and other means. All of these agencies preach the messages of growth and expansion. Farmers are caught on a merry-go-round of change that never stops until … it stops for them.

Governments are happy to follow where the leaders of agribusiness and finance lead them. Most subsidies go to big farms, and those producing crops or livestock that loom large in the food supply and export markets. According to the USDA, about a quarter of all federal farm subsidies go to the largest two percent of farms, and nearly three-quarters to farms in the top ten percent.

Not only that, virtually all subsidies go to support production of five big-ticket crops: wheat, corn, soybeans, rice and cotton. To a very real extent, the true beneficiaries are not the large farmers, but the grain dealers, processors, and marketers who use their products. In effect, government not only undermines opportunities for farmers in other countries through exports of subsidized commodities, but also encourages the introduction into our national food chain of taxpayer-supported foodstuffs.

For example, soybean products are found in about two-thirds of all processed foods. Government subsidies account for 70 percent of the value of American soybeans, making them a cheap and profitable food ingredient, a profit bonanza for the food industry, all at taxpayer expense.

The widespread use of high fructose corn syrup is another example. Never mind the mounting evidence that HFCS is a major cause of obesity and diabetes, it's cheap and that's all that matters. Almost everything you can buy off the supermarket shelves is drenched in the stuff, making more, More, MORE profits for companies like Archer Daniels Midland, Pillsbury, Kraft, and other food behemoths.

Farm Organizations Divide and Conquer

Nominally, farmers do have some clout, through membership in such organizations as the National Corn Growers Association, the American Soybean Association, the National Association of Wheat Growers, the Farm Bureau Federation and many more.

These are essentially special interest advocates that attempt to get the best subsidy deals for their members, often at the expense of other farm groups. It's a constant catfight to get the biggest share from the halls of Congress. Like all lobbying efforts, there's little if any consideration given to the common good, only to the special interest being represented and the Devil take the hindmost. And don't be fooled—the true special interests are usually the corporatist powers behind the curtain, not the farmers who are their supposed clients.

Many of these organizations strive to capture a larger share of the food market for their form of foodstuff. For example, the National Pork Producers Council invests millions to get you to eat more chops, ham and bacon, while the National Cattlemen's Beef Association pushes steaks, roasts and hamburger. The American Dairy Association promotes milk drinking, while the National Corn Growers Association wants us all to consume soft drinks loaded with HFCS.

All this benefits every stage of the food supply chain right to the supermarket shelves. Who really profits? Why, the food processing and marketing corporations of course—not least because a policy of "divide and conquer" keeps farmers at a disadvantage, always dealing from positions of weakness. Instead of giving farmers a unified voice, these groups split them and set them against themselves.

And who pays for all this? You might not want to know, because your head will start to hurt. Well, OK, as you may have suspected farmers pay all or most of the costs of these lobbying and promotional efforts, through membership fees and "check-off" schemes. The producers of our nation's food are like mine workers caught in ever-growing debt to the company store. Any political clout farmers may think they have is mostly of the fantasy kind.

The degree of propagandizing is sometimes startling. For example, the National Corn Growers Association (NCGA) publishes papers such as one in 2009 titled "Ethanol Helps Rural Economies," in which it claims billions of dollars of benefit and nearly half a million jobs from the atrocity of turning subsidized grain into fuel. This is a completely one-sided argument that ignores all the many reasons we explored in an earlier chapter why ethanol from food crops doesn't make economic or environmental sense.

These organizations are essentially propaganda machines, and they don't hesitate to make stuff up to suit their needs. Take another NCGA paper, published in January, 2010, titled "The Future We Seek to Create." It features a chart comparing 2007-08 corn use statistics with supposed 2015-16 and 2020-21 statistics— figures that it notes in the "small print" represent "a vision and not a forecast. It has not been developed with statistical rigor or designed to necessarily balance."[243] To translate this disclaimer into English: the report is without basis in reality and utterly meaningless, part of the program to keep us on the path of "progress."

Most if not all of these groups employ (in truth, are often actually controlled by) Washington lobbyists who hand out political favors in return for that other kind of "pork" that comes not from the farm, but from Congress. Like the corporations that generally control the economy either directly or through their political lackeys, the farm organizations are generally all in favor of "progress," the continued trip down the dead end path. After all, economists tell them this is the right way to think and act, and the association executives in charge get their paychecks no matter what happens to the farmers they supposedly represent.

The Myth that Bigger is Better

Politicians answer to the corporations because they are major sources of campaign money (and perhaps a bit under the

---

[243] Listed at http://ncga.com/publications.

table), so farm programs reflect the desire of agribusinesses to deal with large, specialized operations (or ideally, to directly own or control them). Few government incentives have been offered to help farmers make a living from modest-sized farms or specialty crops. Consolidation is required for power to be wielded at its greatest force.

Despite the constant drumbeat of propaganda for "bigger is better," large farms are not more efficient than small ones except to the extent that they substitute fossil fuels for human labor and attention to the land. Industrial farmers become little more than consumers of industrial products and the drivers of huge and expensive machines that in many cases essentially belong to the local bank as collateral against debt. The art of agriculture has become an assembly line. Farmers have been reduced almost to the level of automatons, doing the bidding of the rich and powerful.

On a per-acre basis, small, intensive, and diversified farms are far more productive because their operators have time to attend to details, to pursue the business of agriculture with more efficient and environmentally friendly methods. Not only that, there's evidence that successful small-scale farmers enjoy greater satisfaction, happiness, and sense of well-being than operators of fully industrialized large-scale farms.

Farm industrialization is a major source of profits for the corporate world. For the farmers themselves, the picture isn't so bright. As mentioned in an earlier chapter, most of the profits from agriculture go into the pockets of the industrial corporations that supply inputs, and to food processors, marketers, and exporters. Anything that's left for the farmer is incidental, crumbs on the table. American wheat growers get only two or three cents from every dollar spent for bread and cereal. Producers of meat, milk and eggs get a bigger share, but costs are higher and margins slim.

Overall, farmers receive less than 20 percent of the food dollar while accepting virtually all the risk. And remember, that 20 percent is gross, not net. A large share of what comes the farmer's way goes right back out to pay for fuel, seed, feed, fertilizer, insurance, interest, and much more. As usual, it's the corporations that are the true beneficiaries. To a casual observer the industrial farmer may appear to be an independent businessman, but in fact he is a kind of modern-day serf, living for the benefit of powerful overlords.

In some years it's not unusual for a large-scale farm operation to barely provide a living to its owners. In fact, many "successful" farm families rely on off-farm employment for much of their living expenses. Most (and sometimes more than all) of their farm income goes to pay for outside expenses.

There's an old joke that expresses the situation. A farmer wins one million dollars in the lottery. Someone asks him what he's going to do with the money. He thinks a moment, then replies: "I think I'll just keep on farming until it's all gone."

And that's not the worst of it, for in times of economic downturn farmers are among the first to take it in the shorts. Agriculture is cyclical, and when a downturn occurs many of the most "progressive" farmers, those who followed the prevailing advice and tried to become bigger by taking on debt, fall into insolvency. When the economic cycle turns down they find themselves on a slippery slope, owing money on loans, needing to purchase expensive inputs just to plant the next crop, experiencing poor market prices and perhaps lousy growing conditions— sometimes even that perfect storm of all the above.

In the atmosphere that industrial agriculture has brought to America, farming can be like living in a war zone. Like it or not, because they are subject to the whims of the markets—markets controlled by the rich and powerful—farmers essentially engage in non-stop economic conflict with their fellow farmers. It's seldom mentioned, but for any given farmer to truly succeed, others must fail, because when everyone harvests a record crop, prices will drop and losses may be suffered by all even in the midst of plenty. Remember the lesson of the "Tragedy of the Commons" that was discussed in an earlier chapter? Today's industrial farmers are in much the same position.

Industrialization has turned farming into a dog-eat-dog competition for survival. Most farmers are by nature kind and generous, and yet in his innermost heart every farmer must hope for the crops of others to fall prey to drought, hail, frost, and plagues of locusts. Only then, when a sufficient number of others fail, can he be assured of a good return from his labor and investment. This is a system of unplanned chaos, created and perpetuated to keep farmers from controlling their own destinies.

No such concerns apply at any other level in the chain that leads from the farm to supermarkets or overseas consumers. The grain dealer, the food processor, the lender, the realtor, the dealers in seed, fuel, fertilizer, machinery, and everything else farmers need to operate—all these reap their profits come rain or come shine, taking their rightful "cut" while farmers are left to swing in the uncertain winds of adversity.

Indeed, when economic troubles arrive it's usually the farmers, not the corporations, who are left hanging. As good old boys might say, "always sucking hind teat." In fact, sometimes agribusiness and food processing corporations profit the most when times are worst for farmers, for example through buying crops at less than the cost of production. Those corporatist entities have no concern for the farmer other than to assure that someone will always be there to till the land and take the risks. To assure that, in difficult times farm lenders wait like vultures circling over a dying mule, ready to foreclose on loans that would never have been made in a rational world and pass the land over to a new generation of serfs.

The Cost of Money Is the Farmer's Bane

Economic cycles periodically wipe out a large number of farm operators. Just such an event happened in the late 1970's and 1980's when inflation fears caused the Federal Reserve to raise the prime rate higher and higher. It was a bonanza for those who owned monetary assets, which could yield as much as 15 percent on secure 30-year treasury notes. Again, it was a chance for the rich to get richer. For those who owed or needed to borrow, it was a disaster—and credit is the lifeblood of industrial farming.

Seeking to "get big" rather than getting out, farmers (not to mention speculators) bid up the value of land. According to a 1977 paper from the Federal Reserve Bank of Chicago:

> Farmland values have exhibited unprecedented increases in recent years. Nationwide, the compound annual rate of increase in farmland prices has been on the order of 16.5 percent during the past five years. The value of an asset appreciating at this rate doubles every four and a half years. If this rate of increase were to persist until the end of the century, land currently valued at $1,000 per acre would be worth $33,535 per acre in the year 2000. If the rate were to drop to one-half the level experienced during the past five years, the value of that same land would rise to "only" $6,192 per acre by the year 2000.[244]

---

[244] Gary L. Benjamin, "Twentieth Century Trends in Farmland Prices," *Economic Perspectives*, a publication of the Federal Reserve Bank of Chicago, May-June, 1977.

During that period of ecstatic land investment farmers were taking on huge amounts of debt, gearing up in response to the government's exhortations to "plant fencerow-to-fencerow." The economy was heating up, and for several years interest rates fluctuated in the mid to high single-digits. A brief run-up in the Fall of 1978 took the prime to 11.75 percent, where it stayed for about five months before beginning to edge upward once more, briefly reaching an unprecedented 20.00 percent on April 2, 1980.

Through the Summer of 1980 the rate receded back to the lower teens, but the worst was still to come. After dropping to an 11 percent rate on July 25, the prime began a steady climb that on December 19 reached 21.50 percent, the highest U.S. prime rate ever recorded. In all, the rate stayed in the range of the upper teens to lower 20 percent for about a year.

Many of the best farmers, those who heeded the call to "get bigger" and took on debt to expand the size and sophistication of their operations, were caught in a fatal bind. Interest on their investments plus operating loans went sky-high. At the same time, their equity fell out of bed as farmland values plummeted. It was the popping of a classic bubble. In Iowa, for example, average farmland prices hit $2147 in 1981, but had dropped to $787 just five years later.[245][246]

Unfortunately, the price they received for their crops didn't

---

[245] Mike Duffy, "*Iowa Trends*," Iowa State University Extension Service, rev. 12/18/09.

[246] It is interesting to note that a new farmland price bubble began to rear its head in 2008. Iowa farmland that sold in 2005 for an average of $2914 per acre rose to $4468, probably in response to the spike in food prices and the hoped-for bonanza of ethanol and biofuel production. Land prices continued to climb. In June, 2015 high quality Iowa cropland was selling for around $7,500 per acre.

follow interest rates to higher levels. U.S. corn prices between 1972 and 2008, with adjustment for inflation, actually declined from a peak in 1974 as the real value of what corn growers received was steadily eroded by inflation. Even during the recent price spike that occurred in 2008, in inflation-adjusted real terms corn prices have never again reached the levels that farmers enjoyed during the early 1970s.

The nominal price of corn has remained remarkably flat for most of the last 50 years or more, while inflation ate away the value from below. In 1960 the price farmers received for a bushel of corn in Illinois averaged $1. Due to inflation, corn would have to sell for more than $8 per bu. today to have equal purchasing power. In fact, at the end of 2015 shelled corn was selling for around $3.50 per bu., back in the nominal range between $2 and $4 that has held constant for many years.

On the surface, the 1970's appeared to be a "good" time for farmers, and yet the statistics for inflation adjusted dollars tells a different story. Consider that while farmers received relatively less value for their corn and other crops, inflation meant that they were spending more for all the inputs they needed. It wasn't just high interest rates that were destroying family farms — the cost of everything was running faster and further ahead of the returns on crop production. During that period of the 1980's, in the U.S. about 2000 family farms went into foreclosure each month.

For those farmers who followed the government and industry lead to take on more debt beginning in the mid-1970s, the falling value of their crops and equity on the one hand, and the rising cost of credit and inputs on the other were like the jaws of a crocodile. Not only did thousands go broke, more than a few took their own lives. A 1991 study by the National Farm Medicine Center, as reported in *he New York Times*,[247] concluded that "[m]ore than 900 male farmers in the Upper Midwest committed suicide in the 1980s, and in some years the incidence of suicide in that group was nearly double the national average for white men."

The study covered the states of Wisconsin, Minnesota, North and South Dakota, and Montana, where a total of 913 male farmers killed themselves during the farm crisis, at a rate that peaked in 1982. The Times report added: "The decade was a particularly stressful time for farmers, with record indebtedness, unstable prices, declining land values and drought. There were thousands of foreclosures and bankruptcies."

And there was more to the story: The report added that between 1980 and 1988 a total of 71 female farmers, 96 farm children, and 177 farm workers also killed themselves in the region. The pattern was repeated from Maine to California. In all, it was a terrible time of stress, failure, and despair for many of our nation's best farmers and their families.

Economic downturns are part of a regular cycle, and nowhere is the impact felt more harshly than on American farms. Each downturn creates new opportunities for the corporate and financial entities as farms are foreclosed and marked down for sale at auction. Buyers may be wealthy investors taking advantage of fire sale prices, or corporations themselves that sometimes hire the former owners to work the land at little better than a minimum wage. All too often, the buyers are part of a new generation heeding the call to farm the land for the benefit of corporate and political interests. The big get bigger. The destruction continues.

---

[247] "Farmer Suicide Rate Swells in 1980's, Study Says," *New York Times*, October 14, 1991.

As we saw in previous chapters, the unscientific art of economics has had a major part in leading farmers down the dead end path of industrialization and "progress." The trip is powered by the continued "development" (destruction) of resources, many of which are non-renewable and most of the rest vastly over-exploited.

In order to justify this depletion, economic models play a kind of shell game, assigning value to debt and credit while denying or minimizing future appreciation of resources and any future costs associated with their development.

Some signs of sanity are occasionally sighted, but these are avis rara indeed in a world hell-bent to convert as much as possible of the natural capital of the planet into short-term benefits and "money" that all too often is merely a fantasy and a fiction, depending for its tangible worth on the "belief" of its holders.

It's a game that only the rich and powerful can win, and it has little purpose other than to yield more riches and power to the players. To this end the entire human race, indeed the planet itself, has been steered on a headlong rush down the dead end path. But the game is nearly over. The end is looming near.

A new form of "real" economics is needed, and soon. Will it emerge in time to save humanity from disaster? This question deserves serious consideration, but sad to say the efforts of the deniers, the exploiters, the politicians, the despoilers, the corporatist powers-that-be are firmly opposed to any such exploration of the hard truths of reality.

Too many inconvenient facts would upset the endless drive for those all-important "values," wealth and power. Those riding in the first class compartment of this Spaceship Earth want to keep things going just the way they are, thank you very much, and since they already control things, they'll probably get their way right up to the yawning gulf that lies ahead.

Sometimes it seems that we may as well ask these other questions: Will pigs grow wings? Will snowballs safely traverse the halls of Hell? And would Diogenes, the ancient Greek who devoted his life to the search for an honest man, find his goal in today's corporate and political landscape of insatiable lust for wealth and power?

Chapter 14

Where Do We Go From Here?

This concludes our investigation of the past and present of industrial agriculture and its place in civilization. We've seen how the destruction of natural resources, the failed model of "progress," runaway corporatism, and the roles of economics and politics have led humanity down a dead end path. There are many conclusions that can be drawn, including these key observations:

- The natural world can remain stable only when there is a balance between the animal and vegetable kingdoms. Agriculture and technology have broken this cycle.

- By making ourselves overlords to Nature, we have engaged in the destruction of the very environment of which we are a part.

- The practice of agriculture, especially in its industrial form, has allowed human population to grow beyond sustainable limits.

- Many nonrenewable resources already are in limited

supply and we are beginning to reach peaks beyond which there are few if any alternatives.

• Our present circumstances are unsustainable and are certain to result in a sharp disconnection with the past as machine technology begins to stutter to a stop due to resource depletion.

• Whatever future lies ahead for humankind will require a return to the balance of Nature, ultimately to a completely sustainable ecology similar to what existed in the earliest stages of agriculture and before.

Peering Beyond the Shadow

We cannot know what is to come, but if anything is certain it's that the past cannot foretell the future. Those who observe the trends of the past, when resources were still in plentiful supply, and blithely project an optimistic line into the future are deluding themselves. The past may be prologue but it predicts nothing.

If you throw a ball into the air do you assume it will continue until it reaches the Moon? No, because like everything else in the Universe the ball is subject to the natural effects of physics. In just this way our present civilization, riding on the back of industrial agriculture and technology, is bound to be brought down by factors such as population overgrowth, resource destruction, climate change, and economic foolishness.

We live in a shaky house of cards built on a foundation of fossil fuels and greed.

One clear fact is that our heirs will not be living in some kind of "Star Trek" future envisioned by extending our machine technology into the future ad infinitum. There are too many reasons to doubt that our existing industrial world can even survive its own excesses, much less lead to ever-higher peaks of "progress."

The direction of human history might be portrayed in the form of a Greek tragedy. The typical plot follows this pattern: In the first act the subject is at the peak of his powers, a shining light among humanity, adored by all. Smitten with his own greatness, he becomes subject to hubris, defined as an excess of ambition or pride. In the second act the tragic figure typically commits cruel acts of arrogance, often in ignorance of the true facts. This displeases the gods (and in our analogy we can give that role to Nature.) In the final act the world comes crashing down around the once-powerful individual, utterly destroying him.

In this analogy our present world is well into the second act of the tragedy, having built a mountain of hubris that is sure to come down upon our collective heads as the real-life drama reaches its end.

What future lies beyond the shadow at the end of the dead end path? We cannot know except to say with assurance that the condition of the human species will be quite different. We can glimpse dim hints of what's to come by considering some basic facts that will bear strongly on the fate of our descendants. It seems essential that any future societies, if humans are to survive at all, must return to a state in balance with Nature. The descent may be quite rapid and if a new balance is to be achieved it must follow these general principles:

- Population must reach equilibrium within the ability of the Earth to sustain it.

- Once that level is reached, human numbers must remain

static or continue to decline. Population growth beyond the natural limits will not be possible when nonrenewable resources are gone.

• Eventually, future technologies must be based entirely on renewable resources or those that are available in essentially unlimited supply. No longer will our descendants live in a world defined by steel, oil, plastics, and all the other products of industrialization.

• A new form of "true" economics must emerge to bring an end to such tragic errors as the pursuit of "progress," rampant materialism, and the endless search for more, More, MORE of everything.

• The world will again become a larger place, one in which no one can fly or drive across oceans and continents at will. In a "true" economy, more or less everything will take place locally.

• Most energy will come from human or animal muscles and natural sources such as wind and flowing water (i.e., sailing vessels, water wheels, and windmills all made from renewable materials). No longer will "work" be done by fossil fuels or through such "solutions" as sophisticated wind generators, nuclear plants, or solar panels that require industrial technology.

• The Earth's ecology must be dramatically adjusted to bring animals, both domestic and wild, back into the cycle of Nature.

• Finally, agriculture itself can and must remain an important part of humanity's future, but in a fully sustainable form. Men and women must become servants of the land, not its masters.

Let's take a final look at these restraints on the future world, and speculate on what a possible sustainable future might be like. It should be realized at the outset that futurists are often wrong, and that unknown factors can cause events to swing wildly in unexpected directions. We're going to explore possibilities, not attempt to read the tea leaves of the future.

Bringing Population Into Balance

As we explored in chapter two, over-population is the 500-pound gorilla in this story. As an unwanted side effect of industrial agriculture—coupled with medical intervention, food aid and other factors—the Earth now has more people than it can sustain. Only through the destruction of irreplaceable natural resources have we been able to reach this state, and those resources are beginning to dwindle.

If we conclude that population cannot continue to grow much further, and that we are at or beyond the level the planet can sustain, then we must accept that human numbers must shrink. There are two ways this can take place, the "hard" way and the "easy" way.[248]

The "hard" way (and it would be really quite difficult and may actually be impossible) would be to proceed through global planning, international cooperation, incentives for reduction in family size, expanded education both for parents and the young, and organized programs to encourage and nurture smaller, less

---

[248] We might be reminded of Paul Ehrlich's "birth rate" and "death rate" solutions, discussed in chapter two.

vulnerable families. In the easy way, population growth would slow and eventually reverse through natural deaths, moving humanity away from the brink of disaster toward a sustainable future.

For generations we have been warned and exhorted to take action against the danger of over-population. There's been some movement toward lower birth rates, but mostly in the rich nations and at least in part due less to responsible action than to the embracing of self-centered lifestyles by adults who choose con–sumerism and self-gratification over parenting.

Still, total numbers have continued to climb at unsustainable rates, and with most of the new births taking place in the poorest parts of the world. There has been little progress in finding an easy way out of this dilemma and little cooperation from any quarter. The easy way is counter to human instincts and perhaps more important, does not suit the aims of "progress" and those who profit or gain power from the irrational model of constant economic growth.

Population control is a "third rail" subject. When rich nations attempt to bring family planning to the world's poor, they're often accused of genocide. Economists declare that growing numbers are necessary to the success of nations. Politicians look to expanding voter rolls to assure their future power. Religious leaders instruct their followers to bring ever more believers into the world.

The hard way just isn't working, at least not yet, and that leaves the alternative—the easy way dictated by reality. In mythical terms, this solution will be delivered by the Four Horsemen of the Apocalypse—War, Famine, Pestilence and Death. That would mean depopulation through disaster, perhaps in successive waves of conflict, starvation, and disease sweeping across the Earth from one region to another as the force of overshoot-and-collapse takes effect. The easy way, in truth, is to do nothing at all and fact the consequences.

It's already happening in some places. Almost everywhere we look in the world there are signs of the Horsemen at work. As we reach or pass Peak Oil and Peak Food, Peak People cannot be far ahead. Add in the possibility of a climate disaster and some even predict the human race will become extinct.[249]

I do not take quite so dim a view of the future for our species. I am convinced that our descendants will pass through a tragic period of collapse and change—perhaps we are already embarked on that difficult journey—but that there may be hope for the survival of our kind. Ironically, one "bright" consideration is my speculation that a significant economic and social breakdown is likely to occur sooner rather than later, before things have gone too terribly wrong. That would skew current projections in a favorable direction.

What if, for example, instead of reaching nine or ten billion by 2050 as the United Nations predicts, the world's population soon begins to fall below its present level through the forces of overshoot-and-collapse?

Tragic, yes, but there would be an upside to such a disaster. If the planet should enter a period of industrial collapse before we literally fall off the end of the dead end path, carbon emissions would slow dramatically, reducing and perhaps even halting global warming. Fewer people would require less food, so famine might also disappear. All the pressures of runaway growth and resource

---

[249] For example, Frank Fenner the Australian scientist who oversaw the elimination of smallpox, said in June, 2010 that "homo sapiens will become extinct, perhaps within 100 years." *The Daily Mail*, on-line edition www.dailymail.co.uk, June 19, 2010.

destruction would be mitigated, giving our descendants a chance to regroup, reconsider, and step away from the path of "progress."

We must presume that it's not too late, if not to avoid the looming disaster, at least to soften it. The hard face of reality is beginning to show itself, and as it does more optimists will move to the pessimist column; more pessimists will become realists. In the face of truly serious disasters, perhaps we'll witness a change in attitudes, with new leaders and organizations emerging to create and direct effective responses.

How likely is that? The record of history provides little encouragement, but it could be possible (many things are). I won't attempt to guess the details of how our race will pass through the looming shadow that lies ahead. It's enough to have described the dangers, and I'll leave it for you to speculate about what our species will experience in the short and medium terms.

Looking Through a Glass Darkly

I want to end this extended essay by speculating on what kind of societies might emerge at the other end of the "shadow" we are about to enter. Let's look further ahead, beyond the chaotic period of uncertainty and trouble, to a possible post-resources time when humans and Nature are once again in balance. That could be relatively soon in an apocalyptic aftermath, or generations from now if a less destructive "solution" is found.

Whenever it occurs, the challenge then will be to contain the tendency to return to a pattern of exponential growth. In a post-resources world that may not be too big a problem, since the limits imposed by Nature will be free to act without the interference of technological methods.

As we saw in chapter two, Malthis understood that these limits would restrain population. His error was in failing to anticipate the potential of industrialization, medicine, and technology to temporarily disable the limits imposed by Nature. In the longer term he will surely be proved correct because his ideas were based on reason and natural law.

In a post-machine era, those limits imposed by Nature will once again come to the forefront, ring-fencing the ability of populations to grow beyond what the planet can support.

It should be understood that even though in post-Peak conditions there will still be large amounts of resources in the ground, they will be the most remote, the most difficult to extract, and the most expensive to obtain. Once the machine technology has collapsed, it will be impossible for later generations to revive it because resources needed to do so will be beyond their reach as the industrial age fades away and a new era emerges.

We've discussed the need for a "true" economy, one that values stability, sustainability, and cooperation. To achieve this will require different ideas and philosophies, perhaps similar to those of Native Americans who revere their places in Nature, as described thus:[250]

> Native American belief systems include many sacred narratives. Such spiritual stories are deeply based in Nature and are rich with the symbolism of seasons, weather, plants, animals, earth, water, sky & fire. The principle of an all embracing, universal and omniscient Great Spirit, a connection to the Earth, diverse creation narratives and collective memories of ancient ancestors are common.

---

[250] Wikipedia, "Native American Mythology."

Old teachings that humans should "go forth and multiply" are no longer useful. They will be less so in the post-Peak world. Such belief systems are rooted in the misguided idea of human ascendancy over the environment, and thus are antithetical to a long-lasting peace with Nature. As crisis and conflict undermine the credibility of the old doctrines and realism begins to gain ascendancy, new philosophies in keeping with "true" economics could start to gain favor and influence. The process would need to continue until anti-Nature ideologies are totally eliminated and replaced by new moral philosophies that return humans to their natural place in harmony with the environment.

Nearly 40 years ago the author Eugene S. Schwartz discussed the changes that would be required for humans to adapt to a new era when the Industrial Age is only a racial memory and the subject of myth. He wrote:

> Technology has been a means to tame the land, the plants, the animals, nature, and finally man. The earth has been impoverished in the process, and man has become alienated and enslaved. To turn aside from a philosophy based on the assumptions of technological civilization, therefore, is not an act of retrogression, a utopian discursion, an escape from reality. It is an act of necessity. By recognizing this necessity man can reach for freedom along new paths. Man can transcend alienation and rediscover himself in the philosophy of post-technological man.[251]

Fortunately, in post-machine times the potential for conflict will be much diminished. War itself has become an industrial process, and in the future it will no longer be possible to deliver

---

[251] Eugene S. Schwartz, "Overskill: The Decline of Technology in Modern Civilization," Ballantine Books, first printing 1972.

mass destruction. We can hope that war will fade away and become an aberration of the past, making the future world a more peaceful one.

When I was a teenager I wrote a short-short story about a time after "the seventh world war" in which a heroine gloried in the power of her new weapon, the latest technological marvel of the 23rd century. At the conclusion her weapon was revealed as a primitive spear. In a post-resources age, nation-states will become impossible things, and these are breeding places of armies and war.

Future citizens of a world in tune with Nature may find pleasure in the study of a stone or observation of the stars, rather than engaging in all the meaningless hustle-and-bustle of the industrial world. We already know that a mind at peace with itself is more likely to be a happy and well-balanced one.

What does more good for a person's spirit:

• Watching a TV sitcom while drinking beer … or sitting quietly in a lovely forest?

• Frantically driving a car along a crowded highway … or walking beside a peaceful mountain stream?

• Running up a credit card on a shopping spree … or making something useful with your hands?

• Consuming a double bacon cheeseburger with fries and a Coke … or picking vegetables from your own garden and gathering eggs for a delicious and wholesome lunch?

All these are examples of the ways in which our lives have become separated from Nature, and how in that process we have devalued the experience of life itself.

David L. Brown

Applying Knowledge

Future societies can benefit from some of the knowledge humanity has gained during the trip down the dead end path, but it must be pruned, focused, and tempered with new-found acceptance of our place in the environment. Ancient knowledge, much of it unfortunately lost in our present time, must be regained or re-invented. Who alive today knows how to craft an elegant spear point from milky flint or dusky obsidian? To start a fire without the aid of matches or a Bic lighter? To live their entire lives with nothing except what was created with their own hands from Nature?

One of the biggest challenges of that distant time will be how to preserve knowledge. When the machine age collapses, our descendants will have no computers, hard drives, memory sticks or DVDs. Eventually not even printed books will survive. As these fade away, what can replace them?

The answers lie in the past, for humans have dealt with this important need before. In a post-technology world, scribes and scholars may emerge once more as they did in the Dark Ages to preserve knowledge the old-fashioned way, through hand-written texts and the power of memory.

Even before the invention of writing ideas were passed from generation to generation through myths and stories created as memory joggers. Ice Age people painted images of their prey on the walls of caves, perhaps to teach the young the arts of the hunter, and to celebrate their connection with Nature. Bards such as Homer preserved history through the recitation of epic poems and legends. Temples were designed and decorated with symbols and inscriptions. Wise sages prepared their successors through teaching and experience. The knowledge of the ages was passed down for thousands of years before we ever dreamed of such things as printing presses, computer chips, and the Internet.

So-called "primitive" people performed amazing feats of architecture and engineering without the use of our machine technology. We still do not know how the Egyptians were able to construct their pyramids five thousand years ago. There are many other examples suggesting that great things can be achieved by a post-industrial society, using renewable resources.

In the far future there is no reason why our descendants will not be able to build new village societies using only the powers of their brains and hands and the materials Nature provides. By necessity those villages will be comfortable, safe, and above all, sustainable.

With the passage of time our vast scientific and technological knowledge will fade away and be replaced by an understanding of the workings of Nature—knowledge of the plants that can be used for medicine; the skills to weave a blanket from wool or camel's hair; the ability to erect fine buildings from stone and wood; understanding of the secrets of the animal world; plus all the other lore that our distant forebears knew and practiced. Some of our present knowledge may help us regain those skills, but it will not be easy.

In the far distant future there is no doubt but that our era will be looked upon as a mythical lost age, not unlike the legend of Atlantis. Imagine the members of a future community, perhaps sitting in a circle as an itinerant storyteller relates the legends of our time from the mists of history. Will those listeners be able to believe what they hear? Almost certainly not, for it will seem far too unlikely to have actually happened. (Except, of course, that we know it is.) The stories will be passed off as mere figments of the

David L. Brown

storyteller's imagination. No one, they will think, could be so foolish as to build a civilization that would destroy itself.

Make This World a Promised Land

In our imagined distant future when the time of the shadow has been left behind, the simple pleasures of everyday life will be the essence of human existence. No longer will our descendants live in a fantasy world set apart from the natural one—in cities where no one can ever see the Milky Way; in dwellings artificially cooled in summer and heated in winter to divide us from the seasons; in lives spent performing meaningless tasks that leave "retirement" as the ultimate goal of existence. Remembering Wordsworth's statement, "The world is too much with us," many might admit that such lives are of questionable value.

If our descendants can make the transition to a balanced, "true" economy in the distant future, everyone will be fully engaged with reality. It will be a world in which work has real meaning and the calluses to prove it. Food will be something one produces with sweat and through cooperation with fellows.

Beyond that, family and community relationships will be close-knit and personal. Rather than high-rise apartments or sub-divisions, most people will live in villages, small communities where everyone knows everyone else and mutual trust lies at the heart of each interaction. Life will be something of value, something worth enjoying every day.

This may seem like an alien concept to those of us living in a dispersed, impersonal society where many of us do not even know our neighbors—but until recently that is just how most of our ancestors lived, and in developing regions many still do. The village was and will again be the center of life for most people. One

312

might occasionally visit a nearby town for courting or trade, but for most the wider world beyond will be merely the subject of rumor and speculation. News will come from the occasional traveler or storyteller from strange places.

Those who live in post-Peak times, beyond the shadow that lies at the end of the dead end path, will need to gather themselves into just such social units. Let us call them tribes. The current motto: "Think Globally, Act Locally" will be reduced to the last two words. Everything will be local. Globalization will be a forgotten and discredited idea, a delusion in the name of "progress" and conceivable only through unsustainable "development."

We do not live so far from that simple world as we think. When I was growing up in central Missouri in the late 1940s and early 1950s, for five years I attended classes in a one-room schoolhouse.[252] One teacher, eight grades, eventually more than 80 pupils crowded into a space about 35 feet square. Outside was a cistern with a hand pump and drinking cup. Two smelly his-and-hers outhouses were located up the hill, no treat when the snow was drifted deep and the wind chill was zero or below.

Many of the pupils were from a single family, the Bakers. The father drove a delivery truck for the local coal company and at that time the family—eventually 14 children plus two adults—lived in a 20-by-20 ft. war surplus tent pitched near the bank of a stream. The Baker kids hiked to school about two miles cross-country through fields and woods, rain, shine, or snow.

In the center of their tent was an old pot-bellied stove for

---

[252] The Keene School near Columbia, MO was built in 1898 and classes were held there through 1953, the year after I left. The brick building was converted to a residence with a second floor added in the original high ceiling. It is a registered historic landmark.

cooking and heat. The floor was Nature's own earth, and ranged around the outside walls were a number of surplus double-decker bunk beds with a collection of Army blankets. There were a few footlockers and trunks to contain all that the Bakers owned.

When she was in about the fifth grade one of the Baker girls wrote a theme in which she described how she and her sisters bathed in the winter. It involved going down to the creek with an axe and cutting a hole in the ice. Their water for cooking and drinking came from the same source.

The oldest Baker boy always came to school with a rifle or shotgun, concealed during class in a hollow tree across the gravel road from the school. It was his job to hunt on the way to and from school to help provide food for the family. He was charged with bringing something for the pot for each shot-shell or cartridge his father doled out, whether squirrels, rabbits, raccoons, opossums or even bullfrogs.

He showed me the art of catching fish with his bare hands. Although only a boy, he was a natural woodsman and dead-eye shot who once tracked and killed a wildcat in the dark of night, earning a $35 bounty for its ears—a small fortune in those days, especially for a family like theirs.

The Baker family might fit fairly easily into the post-resources world, but for most of us their lifestyle is inconceivable. Could we even imagine living in such a way? And yet, I remember the Baker clan as basically a happy bunch, unashamed of their status, each standing up for the others. They were a tribe. Setting aside their contribution to over-population, perhaps they had already discovered the secrets of the good life, one that is simple and close to Nature.

The Future Must Be Sustainable

As we have seen, nearly all of the non-renewable resources on which the industrial world depends are in decline. Eventually, none of these will be available to our presumed descendants living a natural existence in harmony with the environment. Those missing resources will include almost everything that we take for granted today, including iron, copper, zinc, tin, indeed almost the entire Periodic Table of the Elements. Only those things that are renewable or in essentially inexhaustible supply will be available for the post-industrial technology of such a future world.

That suggests that human society will need to draw back not only from our present age, but even beyond the Iron Age, the Bronze Age, and the Copper Age. Must we revert all the way to the Stone Age? Perhaps not entirely, because we have gained a lot of knowledge and can learn more. As we've seen, there are many possibilities for renewable, low-tech crafts; the application of plant and animal selection; ways to build and make useful things from renewable materials; and methods of protecting rather than destroying the environment. It will be a different path, but not necessarily a downward one.

Diversity will be a necessary feature of that time, with each section of the patchwork quilt of humanity adapting itself to its particular ecology, its opportunities, and its spiritual needs. If you doubt that there are numerous ways to make a living within Nature, consider the Inuit living in a world of ice and snow, the Pygmies of the unforgiving Kalahari Desert, or the Yanomami who live in the Amazonian rain forest.

Our descendants in that distant, post-industrial time will have learned once more to work with wood, stone, sand, and clay; to weave plant and animal fibers into cloth; and to build sturdy and comfortable dwellings. They will have created forms of agriculture that conform to the "true" economics of sustainability. Some will have become herdsmen and dairymen to let animals return to the cycle of nature, harvesting the grass from hillsides and providing their manure to fertilize and renew crop fields. Like humans, animals must regain their natural place in Nature, never to be separated from the land as they are today. The cycle between the plant and animal kingdoms must be repaired and made whole.

Those future humans will have learned the skills of the gardener and herbalist. They will have reinvented and improved the ways of crafting tools from wood, flint, and obsidian as our "primitive" (but oh-so-clever) ancestors did. They will teach their children to conserve all that is in the natural world—to prevent the loss or destruction of precious soil; to work with animals and harvest the energy of the wind and flowing streams; to respect the environment and all that is within it.

In mythical terms, such a scenario would represent a return to the Garden of Eden. If she could speak to us directly, the Creator would approve. Her secret name is Nature. She is the Mother of us all.

# # # # #

Bibliography

The following are books that I read or referred to in the preparation of *Dead End Path*. All are in my personal library. In addition to these I have read and collected many other sources too numerous to mention, including articles from *Science*, *New Scientist*, *Scientific American*, and *The Economist*, plus many papers and reports discovered on the World Wide Web. Some of these are quoted and referenced in footnotes. The volumes listed below are provided here in recognition of their roles as general sources of background information. — DLB

A

Adams, Barbara Berst, *Micro Eco-Farming*, New World Publishing, 2004.

B

Barlow, Maude, *Blue Covenant: The Global Water Crisis and the Coming Battle for the Right to Water*, The New Press, 2007.
Barney, Gerald O. (study director), *The Global 2000 Report to the President*, Penguin Books, 1982.
Battan, Louis J., *Harvesting the Clouds: Advances in Weather Modification*, Doubleday & Co., 1969.

Benarde, Melvin A., *Our Precarious Habitat*, W. W. Norton & Co., 1970.

Berry, Wendell, *The Unsettling of America: Culture and Agriculture*, Sierra Club Books, 1977.

Borgstrom, Georg, *The Hungry Planet*, The Macmillan Company, 1965.

Bradford, Travis, *Solar Revolution*, The MIT Press, 2006.

Breuil, H. and Lantier, R., *The Men of the Old Stone Age*, George G. Harrap & Co., 1965.

Bronowski, J., *The Ascent of Man*, Little, Brown & Co., 1973.

Brown, Harrison, *The Challenge of Man's Future*, Viking Press, 1954.

Brown, Lester, *Outgrowing the Earth*, W. W. Norton & Co., 2004.

— *Plan B 3.0*, W. W. Norton & Co., 2008.

— *Eco-Economy: Building an Economy for the Earth*, W. W. Norton & Co., 2001.

Burlingame, Roger, *Machines the Built America*, Signet Books, 1953.

Burroughs, William J., *Climate Change in Prehistory*, Cambridge University Press, 2005

C

Catton, William R. Jr., *Overshoot*, University of Illinois Press, 1980.

Clarke, Arthur C., *Profiles of the Future*, Holt, Rinehart & Winston, 1984.

Charles, Daniel, *Lords of the Harvest: Biotech, Big Money and the Future of Food*, Perseus Publishing, 2001.

Clay, Jason, *World Agriculture and the Environment*, Island Press, 2004.

Childe, V. Gordon, *Man Makes Himself*, New American Library, 1951.

Commoner, Barry, *The Poverty of Power: Energy and the Economic Crisis*, Alfred A. Knopf, 1976.

Conkin, Paul K., *A Revolution Down on the Farm*, The University Press of Kentucky, 2008.

Cohen, Joel E., *How Many People Can the Earth Support?*, W. W. Norton & Co., 1995.

Cox, John D., *Climate Crash*, Joseph Henry Press, 2005.

D

Darwin, Charles, *The Autobiography of Charles Darwin*, new edition, W. W. Norton & Co., 1958.

Deffeyes, Kenneth S., *Beyond Oil: The View from Hubbert's Peak*, Hill and Wang, 2005.

Diamond, Jared, *Collapse: How Societies Choose to Fail or Succeed*, Viking Press, 2005.

E

Ehrlich, Paul, *The Population Bomb*, Buccaneer Books, 1968.

Ehrlich, Paul and Ehrlich, Anne, *One With Nineveh: Politics, Consumption and the Human Future*, Island Press, 2004.

— *Population/Resources/Environment: Issues in Human Ecology*, W. H. Freeman & Co., 1970.

— *The Population Explosion*, Simon and Schuster, 1990.

F

Fagan, Brian, *Floods, Famines and Emperors: El Nino and the Fate of Civilizations*, Basic Books, 1999.

— *The Long Summer: How Climate Changed Civilization*, Basic Books, 2004.

Flannery, Tim, *The Weather Makers*, Atlantic Monthly Press, 2005.

G

Galbraith, John Kenneth, *The New Industrial State*, Houghton Mifflin Co., 1967

Garrett, Laurie, *The Coming Plague: Newly Emerging Diseases In a World Out of Balance*, Farrar Straus and Giroux, 1994.

Glennon, Robert, *Water Follies: Groundwater Pumping and the Fate of America's Fresh Waters*, Island Press, 2002.

Gorbachev, Mikhail, *Manifesto for the Earth*, Clairview Books, 2006.

Gore, Al, *Earth in the Balance*, Houghton Mifflin, 1992.

— *An Inconvenient Truth*, Rodale, Inc., 2006.

— *Our Choice*, Rodale, Inc., 2009.

Greer, John Michael, *The Ecotechnic Future: Envisioning a Post-Peak World*, New Society Publishers, 2009.

Grigg, D.B., *The Agricultural Systems of the World*, Cambridge University Press, 1974.

H

Handler, Philip (editor), *Biology and the Future of Man*,
Oxford University Press, 1965

Hawken, Paul et al., *Natural Capitalism: Creating the
Next Industrial Revolution*, Little, Brown, 1999.

Hayek, F. A., *The Road to Serfdom*, The University of
Chicago Press, 2007.

Hazlett, Henry, *Economics in One Lesson*, Three Rivers
Press, rev. ed. 1978.

Heinberg, Richard, *Peak Everything*, New Society Publishers, 2007.

— *Power Down: Options and Actions for a Post-Carbon World*, New
Society Publishers, 2004.

— *A New Covenant with Nature*, Quest Books, 1996.

Heppenheimer, T.A., *Colonies in Space*, Stackpole Books, 1977.

Hieronymus, Thomas A., *Economics of Futures Trading*, Commodity
Research Bureau, Inc., 1971.

Hodges, Henry, *Technology in the Ancient World*, Alfred A. Knopf, 1970.

Houghton, John, *Global Warming: The Complete Briefing*,
3rd Ed., Cambridge University Press, 2004.

I

Ikerd, John, *Sustainable Capitalism*, Kumarian Press, 2005.

— *A Return to Common Sense*, R. T. Edwards, Inc., 2007.

— *Small Farms Are Real Farms*, Acres U.S.A., 2008.

— *Crisis & Opportunity*, University of Nebraska Press, 2008.

Immerwahr, George E., *World Population Growth*,
Peanut Butter Publishing, 1995.

J

Jensen, Derrick, *Endgame: The Problem of Civilization*,
Seven Stories Press, 2006.

Jensen, Derrick and McBay, Aric, *What We Leave Behind*,
Seven Stories Press, 2009.

K

Keith, Lierre, *The Vegetarian Myth*, Flashpoint Press, 2009.

Kelsey, Darwin P. (editor), *Farming in the New Nation, 1790-1840,* The Agricultural History Society, 1972

Kimbrell, Andrew (editor), *The Fatal Harvest Reader: The Tragedy of Industrial Agriculture,* Island Press, 2002.

Kolbert, Elizabeth, *Field Notes from a Catastrophe: Man, Nature and Climate Change,* Bloomsbury, 2006.

Kramer, Mark, *Three Farms,* Bantam Books, 1981.

Kunstler, James Howard, *The Long Emergency,* Atlantic Monthly Press, 2005.

L

Landels, J. G., *Engineering in the Ancient World,* University of California Press, 1978.

Lappé, Marc, *Broken Code: The Exploitation of DNA,* Sierra Club Books, 1984.

Leopold, Aldo, *A Sand County Almanac,* Oxford University Press, 1949.

Linden, Eugene, *The Winds of Change,* Simon & Schuster, 2006.

Lovelock, James, *The Revenge of Gaia,* Basic Books, 2006

— *The Vanishing Face of Gaia,* Basic Books, 2009.

Lynas, Mark, *Six Degrees,* National Geographic Society, 2008.

M

MacGowan, Kenneth and Hester, Joseph A. Jr., *Early Man in the New World,* Doubleday Anchor Books, 1962.

MacKay, Charles, *Extraordinary Popular Delusions and the Madness of Crowds,* originally published 1841. Facsimile Edition, L. C. Page & Co., 1932.

Malthus, Thomas R., *An Essay on the Principle of Population,* Oxford University Press, World's Classics Edition, 1993.

Manning, Richard, *Grassland,* Penguin Books, 1995.

— *Against the Grain: How Agriculture Has Hijacked Civilization,* North Point Press, 2004.

— *Food's Frontier: The Next Green Revolution,* The University of California Press, 2000.

Mazoyer, Marcel and Roudart, Laurence, *A History of World Agriculture from the Neolithic Age to the Current Crisis*, Monthly Review Press, 2006.

McKibben, Bill, *Deep Economy: The Wealth of Communities and the Durable Future*, Henry Holt & Co., 2007.

McNeill, *Plagues and Peoples*, Anchor Press/Doubleday, 1976.

Meadows, Donalia H. et al., *The Limits to Growth*, 2nd ed., Universe Books, 1974.

— *Beyond the Limits*, Chelsea Green Publishing Co., 1992.

— *Limits to Growth: The 30-Year Update*, Chelsea Green Publishing Co., 2004.

Medawar, P.E., *The Future of Man*, Mentor Books, 1959.

Muir, John, *Wilderness Essays*, Peregrine Smith Books, 1980.

Mumford, Lewis, *Technics and Human Development*, Harcourt Brace Jovanovich, 1967.

— *The Pentagon of Power*, Harcourt Brace Jovanovich, 1964.

N

Naisbitt, John, *Megatrends*, Warner Books, 1984.

Naisbitt, John and Aburdene, Patricia, *Megatrends 2000*, Avon Books, 1990.

O

Osborn, Fairfield, *Our Plundered Planet*, Little, Brown & Co., 1948.

P

Pearce, Fred, *When the Rivers Run Dry*, Beacon Press, 2006.

— *With Speed and Violence*, Beacon Press, 2007.

Pfeiffer, Dale Allen, *Eating Fossil Fuels: Oil, Food and the Coming Crisis in Agriculture*, New Society Publishers, 2006.

Pfeiffer, John E., *The Creative Explosion: An Inquiry into the Origins of Art and Religion*, Harper & Row, 1982.

— *The Emergence of Man*, Harper & Row, 1969.

— The Emergence of Society, McGraw Hill, 1977.

Pyke, Magnus, *Man and Food*, McGraw Hill, 1970.

R

Rand, Ayn, *For the New Intellectual*, Random House, 1961.

Reich, Charles A., *The Greening of America*, Bantam Books, 1971.

Reisner, Marc, *Cadillac Desert: The American West and Its Disappearing Water*, Penguin Books, 1987.

Roberts, Paul, *The End of Food*, Mariner Books, 2008.

Ruddiman, William F., *Plows, Plagues & Petroleum*, Princeton University Press, 2005.

Ruppert, Michael C., *Confronting Collapse*, Chelsea Green Publishing, 2009.

S

Sampson, R. Neil, *Farmland or Wasteland: A Time to Choose*, Rodale, Inc., 1981.

Samuelson, Robert J., *The Great Inflation and Its Aftermath*, Random House, 2008.

Schumacher, E. F., *Small Is Beautiful: Economics As If People Mattered*, Harper Perennial, reissue 1989.

Schlebecker, John T., *Whereby We Thrive: A History of American Farming, 1607-1972*, The Iowa State University Press, 1975.

Schneider, Stephen H. and Randi Londer, *The Coevolution of Climate & Life*, Sierra Club Books, 1984.

Schwartz, Eugene S., *Overskill: The Decline of Technology in Modern Civilization*, Ballantine Books, 1971.

Shepherd, Geoffrey S. et al., *Marketing Farm Products*, The Iowa State University Press, 6th ed. 1976.

Shideler, James H. (editor), *Agriculture in the Development of the Far West*, The Agricultural History Society, 1975.

Shiva, Vandana, *Stolen Harvest: The Hijacking of the Global Food Supply*, South End Press, 2000.

Simmons, Matthew R., *Twilight in the Desert*, John Wiley & Sons, Inc., 2005.

Skinner, B. F., *Beyond Freedom and Dignity*, Bantam Books, 1971.

Soddy, Frederick, *Wealth, Virtual Wealth and Debt*, orig. ed. 1923, new printing CPA Book Publisher, 1983

Suzuki, David, *The David Suzuki Reader*, Greystone Books, 2003.

Suzuki, David and Dressel, Holly, *Good News for a Change*, Greystone Books, 2002.

T

Tainter, Joseph A., *The Collapse of Complex Societies*, Cambridge University Press, 1988.

Thoreau, Henry David, *The Natural History Essays*, Peregrine Smith Books,1980.

Toffler, Alvin, *Future Shock*, Bantam Books, 1970.

— *The Third Wave*, William Morrow and Company, Inc., 1980.

— *Powershift*, Bantam Books, 1990.

W

Ward, Barbara, and Dubos, Rene, *Only One Earth: The Care and Maintenance of a Small Planet*, W. W. Norton & Co., 1972.

Wessel, James, *Trading the Future*, Institute for Food and Development Policy, 1983.

Whitaker, James W. (editor), *Farming in the Midwest 1840-1900*, The Agricultural History Society, 1974.

Winne, Mark, *Closing the Food Gap*, Beacon Press, 2008.

Woodward, William E., *The Way Our People Lived: An Intimate American History*, Washington Square Press, 1965.

World Commission on Environment and Development, *Our Common Future*, Oxford University Press, 1987.

Worldwatch Institute, *2010 State of the World: Transforming Cultures*, W. W. Norton & Co., 2010.

— *2007 State of the World: Our Urban Future*, W. W. Norton & Co., 2007.

— *Vital Signs 2006-2007: The Trends that Are Shaping Our Future*, W. W. Norton & Co., 2006.

Acknowledgements

My thanks to my friends who encouraged and helped me in the creation of *Dead End Path* and this second revised edition. They include Val Germann, Alexandra Dell'Amore, Kim Sherwood, Barbara Deputy, Steven Hughes, David Ponton, and Cherie Major plus literally thousands of individuals I have met and worked with in the course of my life, each of whom brought something to my experience and understanding.

www.ingramcontent.com/pod-product-compliance
Lightning Source LLC
Chambersburg PA
CBHW060004210326
41520CB00009B/819